Advances in Intelligent Systems and Computing

Volume 1007

The series “Advances in Intelligent Systems and Computing” contains publications on theory, applications, and design methods of Intelligent Systems and Intelligent Computing. Virtually all disciplines such as engineering, natural sciences, computer and information science, ICT, economics, business, e-commerce, environment, healthcare, life science are covered. The list of topics spans all the areas of modern intelligent systems and computing such as: computational intelligence, soft computing including neural networks, fuzzy systems, evolutionary computing and the fusion of these paradigms, social intelligence, ambient intelligence, computational neuroscience, artificial life, virtual worlds and society, cognitive science and systems, Perception and Vision, DNA and immune based systems, self-organizing and adaptive systems, e-Learning and teaching, human-centered and human-centric computing, recommender systems, intelligent control, robotics and mechatronics including human-machine teaming, knowledge-based paradigms, learning paradigms, machine ethics, intelligent data analysis, knowledge management, intelligent agents, intelligent decision making and support, intelligent network security, trust management, interactive entertainment, Web intelligence and multimedia.

The publications within “Advances in Intelligent Systems and Computing” are primarily proceedings of important conferences, symposia and congresses. They cover significant recent developments in the field, both of a foundational and applicable character. An important characteristic feature of the series is the short publication time and world-wide distribution. This permits a rapid and broad dissemination of research results.

**** Indexing: The books of this series are submitted to ISI Proceedings, EI-Compendex, DBLP, SCOPUS, Google Scholar and Springerlink ****

More information about this series at http://www.springer.com/series/11156

Rosella Gennari · Pierpaolo Vittorini ·
Fernando De la Prieta · Tania Di Mascio ·
Marco Temperini · Ricardo Azambuja Silveira ·
Demetrio Arturo Ovalle Carranza
Editors

Methodologies and Intelligent Systems for Technology Enhanced Learning, 9th International Conference

Editors
Rosella Gennari
Computer Science Faculty
Free University of Bozen-Bolzano
Bolzano, Italy

Fernando De la Prieta
Department of Computer Science
and Automation Control
University of Salamanca
Salamanca, Salamanca, Spain

Marco Temperini
Dipartimento di Ingegneria Informatica,
Automatica e Gestionale
Sapienza Università di Roma
Rome, Italy

Demetrio Arturo Ovalle Carranza
Universidad Nacional de Colombia
Medellin, Colombia

Pierpaolo Vittorini
Department of Life, Health
and Environmental Sciences
University of L'Aquila
L'Aquila, Italy

Tania Di Mascio
Department of Information Engineering,
Computer Science and Mathematics
University of L'Aquila
L'Aquila, Italy

Ricardo Azambuja Silveira
Department of Computer Science
and Statistics
Federal University of Santa Catarina
Florianópolis, Brazil

ISSN 2194-5357 ISSN 2194-5365 (electronic)
Advances in Intelligent Systems and Computing
ISBN 978-3-030-23989-3 ISBN 978-3-030-23990-9 (eBook)
https://doi.org/10.1007/978-3-030-23990-9

This Springer imprint is published by the registered company Springer Nature Switzerland AG
The registered company address is: Gewerbestrasse 11, 6330 Cham, Switzerland

Preface

Education is the cornerstone of any society, and it serves as one of the foundations for many of its social values and characteristics. Different methodologies and intelligent technologies are employed for creating Technology Enhanced Learning (TEL) solutions. Solutions are innovative when they are rooted in artificial intelligence, deployed as stand-alone solutions or inter-connected to others. They target not only cognitive processes but also motivational, personality, or emotional factors. In particular, recommendation mechanisms enable us tailoring learning to different contexts and people, e.g., by considering their personality. The use of learning analytics also helps us augment learning opportunities for learners and educators alike; e.g., learning analytics can support self-regulated learning or adaptation of the learning material. Besides technologies, methods help create novel TEL opportunities. Methods come from different fields, such as education, psychology or medicine, and from diverse communities where people collaborate, such as *making communities* and *participatory design communities*. Methods and technologies are also used to investigate and enhance learning for "fragile users", like children, elderly people, or people with special needs.

Both the 9th edition of this conference and its related workshops (i.e., Nursing, Tel4creativity, and TELAssess) contribute to novel research in TEL and expands the topics of the previous editions. The MIS4TEL 2019 papers discuss how diverse methods or technologies are employed to create novel approaches to TEL, valuable TEL experiences, or innovative TEL solutions, taking a critical stance, and promoting innovation.

This volume presents all papers that were accepted for the main track of MIS4TEL 2019, while the workshop papers will be published in a different volume. All underwent a peer-review selection: each paper was assessed by at least two different reviewers, from an international panel composed of about 50 members of 15 countries. The program of MIS4TEL counted 20 contributions from several countries, such as Brazil, Colombia, Germany, Greece, Italy, Mexico, Oman, Romania, Russia, Slovakia, Spain, and Sweden. The quality of submissions was on average good, with an acceptance rate of approximately 70%.

We thank the sponsors (IEEE Systems Man and Cybernetics Society - Spain Section Chapter, IEEE Spain Section, IBM, Indra, Viewnext, Global exchange, AEPIA, AIR institute, and APPIA), the members of the Local Organization team, and the Program Committee members for their hard work, which was essential for the success of MIS4TEL'19.

Rosella Gennari
Pierpaolo Vittorini
Fernando De la Prieta
Tania Di Mascio
Marco Temperini
Ricardo Azambuja Silveira
Demetrio Arturo Ovalle Carranza

Organization of MIS4TEL 2019

http://www.mis4tel-conference.net/

General Chair

Rosella Gennari	Free University of Bozen-Bolzano, Italy

Technical Program Chair

Pierpaolo Vittorini	University of L'Aquila, Italy

Paper Co-chairs

Tania Di Mascio	University of L'Aquila, Italy
Fernando De la Prieta	University of Salamanca, Spain
Ricardo Azambuja Silveira	Universidade Federal de Santa Catarina, Brazil
Marco Temperini	Sapienza University, Rome, Italy

Proceedings Chairs

Ana Belén Gil	University of Salamanca, Spain
Fernando De la Prieta	University of Salamanca, Spain

Publicity Chair

Demetrio Arturo Ovalle Carranza	National University of Colombia, Colombia

Workshop Chair

Elvira Popescu	University of Craiova, România

Program Committee

Sara Rodríguez	University of Salamanca, Spain
Juan M. Alberola	Universitat Politècnica de València, Spain
Juan M. Santos	University of Vigo, Spain
Sonia Verdugo-Castro	Universidad de Salamanca, Spain
Cecilia Giuffra	UFSC, Brazil
Sérgio Gonçalves	University of Minho, Portugal
Vincenza Cofini	University of L'Aquila, Italy
Alessandra Melonio	Free University of Bozen-Bolzano, Italy
Angélica González Arrieta	Universidad de Salamanca, Spain
Vicente Julian	Universitat Politècnica de València, Spain
Samuel González-López	Technological Institute of Nogales, Mexico
Orazio Miglino	NAC Lab, University of Naples "Federico II" and LARAL, Institute of Cognitive Sciences and Technologies, CNR, Italy
Paulo Novais	University of Minho, Portugal
Jorge Gomez-Sanz	Universidad Complutense de Madrid, Spain
Ana Belén Gil González	University of Salamanca, Spain
Ana Faria	ISEP, Portugal
Fridolin Wild	Oxford Brookes University, UK
Ana Almeida	ISEP-IPP, Portugal
Tiago Primo	Federal University of Pelotas, Brazil
Margarida Figueiredo	Universidade de Évora, Portugal
Juan-José Mena-Marcos	University of Salamanca, Spain
Antonio J. Sierra	University of Seville, Spain
Henrique Vicente	University of Évora, Portugal
Diogo Cortiz	Pontificia Universidade Católica de São Paulo, Brazil
Carlos Pereira	ISEC, Portugal
Marcelo Milrad	Linnaeus University, Sweden
Florentino Fdez-Riverola	University of Vigo, Spain
Elvira Popescu	University of Craiova, Romania
Jose Neves	University of Minho, Portugal
Laura Tarantino	Università dell'Aquila, Italy
Ricardo Azambuja Silveira	Universidade Federal de Santa Catarina, Brazil
Davide Carneiro	Polytechnic Institute of Porto, Portugal
Carolina Schmitt Nunes	Universidade Federal de Santa Catarina, Brazil
Dalila Duraes	Department of Artificial Intelligence, Technical University of Madrid, Madrid, Spain

Mauro Caporuscio	Linnaeus University, Sweden
Besim Mustafa	Edge Hill University, UK
Katherine Maillet	Institut Mines-Télécom, Télécom Ecole de Management, France
Giovanni De Gasperis	DISIM, Univ. L'Aquila, Italy
Gerlane R. F. Perrier	Universidade Federal Rural de Pernambuco, Brazil

Local Organising Committee

Juan Manuel Corchado Rodríguez	University of Salamanca, Spain and AIR Institute, Spain
Fernando De la Prieta	University of Salamanca, Spain
Sara Rodríguez González	University of Salamanca, Spain
Sonsoles Pérez Gómez	University of Salamanca, Spain
Benjamín Arias Pérez	University of Salamanca, Spain
Javier Prieto Tejedor	University of Salamanca, Spain and AIR Institute, Spain
Pablo Chamoso Santos	University of Salamanca, Spain
Amin Shokri Gazafroudi	University of Salamanca, Spain
Alfonso González Briones	University of Salamanca, Spain and AIR Institute, Spain
José Antonio Castellanos	University of Salamanca, Spain
Yeray Mezquita Martín	University of Salamanca, Spain
Enrique Goyenechea	University of Salamanca, Spain
Javier J. Martín Limorti	University of Salamanca, Spain
Alberto Rivas Camacho	University of Salamanca, Spain
Ines Sitton Candanedo	University of Salamanca, Spain
Daniel López Sánchez	University of Salamanca, Spain
Elena Hernández Nieves	University of Salamanca, Spain
Beatriz Bellido	University of Salamanca, Spain
María Alonso	University of Salamanca, Spain
Diego Valdeolmillos	University of Salamanca, Spain and AIR Institute, Spain
Roberto Casado Vara	University of Salamanca, Spain
Sergio Marquez	University of Salamanca, Spain
Guillermo Hernández González	University of Salamanca, Spain
Mehmet Ozturk	University of Salamanca, Spain
Luis Carlos Martínez de Iturrate	University of Salamanca, Spain and AIR Institute, Spain
Ricardo S. Alonso Rincón	University of Salamanca, Spain
Javier Parra	University of Salamanca, Spain
Niloufar Shoeibi	University of Salamanca, Spain
Zakieh Alizadeh-Sani	University of Salamanca, Spain

Belén Pérez Lancho	University of Salamanca, Spain
Ana Belén Gil González	University of Salamanca, Spain
Ana De Luis Reboredo	University of Salamanca, Spain
Emilio Santiago Corchado Rodríguez	University of Salamanca, Spain
Angel Luis Sánchez Lázaro	University of Salamanca, Spain

Contents

Towards an Annotation System for Collaborative Peer Review

Sebastian Mader(✉) and François Bry

Institute for Informatics,
Ludwig Maximilian University of Munich, Munich, Germany
sebastian.mader@ifi.lmu.de

Abstract. Peers providing feedback on their peers' work is called peer review which has been shown to have beneficial effects on students' learning. This article presents a novel approach to peer review where reviewers, reviewees, and lecturers alike have access to the same documents, and reviews are created in form of annotations which are automatically shared among all users, can be up- or downvoted, and can themselves be commented on. Working on a same document and seeing annotations immediately after they have been created enables various forms of collaboration among learners: Between reviewers who can agree or disagree with a review by up- or downvoting or comment on the review thus providing further insight, between reviewers and reviewees by allowing the reviewees to inquire about reviews and reviewers to provide clarifications and justifications, and between lecturers and reviewers, when lecturers take the role of reviewers and up- or downvote and comment on reviews. The contributions of this article are twofold: First, a report on the conception and implementation of a collaborative annotation environment that supports the collaborative peer review described in this article, and second, an analysis of the communication that happened during collaborative peer review in three courses pointing to the approach's effectiveness but uncovering problems of the collaborative annotation system as well.

Keywords: Peer review · Computer-supported collaborative learning · Virtual learning environments

1 Introduction

Peer review is an arrangement in which peers provide feedback on or marks for their peers' work which has been shown to have beneficial effects on reviewers' and reviewees' learning alike [15]. Among the benefits of peer review are its reflective nature, students spending more time on the task, as well as students getting more and faster feedback [15]. At times where large classes are becoming more and more common, peer review can be used to lessen the lecturers' workload [9] allowing them to focus on students that require their personal feedback.

R. Gennari et al. (Eds.): MIS4TEL 2019, AISC 1007, pp. 1–10, 2020.
https://doi.org/10.1007/978-3-030-23990-9_1

Traditionally in document-based peer review, reviewees' documents are assigned to one or more reviewers who then work out their reviews independently from other reviewers which are afterwards delivered to the reviewees. This strict sequencing of phases has a number of disadvantages: If several reviewers are involved, it is likely that the same work will be done repeatedly with reviewers marking mostly the same parts and only differing in a few comments. Reviewees can not easily ask questions about ambiguities or justify their choices because the review phase is already completed and reviewers are often no longer available.

To blur the various phases of traditional peer review, this article introduces technology-supported collaborative peer review where the aforementioned phases are running concurrently. Reviewers, reviewees, and lecturers alike have access to the same documents, and reviews are created in form of annotations which are automatically shared among all users, can be up- or downvoted, and can themselves be commented on. Working on a same document and seeing annotations immediately after they have been created enables various forms of collaboration among learners: Between reviewers who can agree or disagree with a review by up- or downvoting or comment on the review thus providing further insight, between reviewers and reviewee by allowing the reviewee to inquire about reviews and reviewers to provide clarifications and justifications, and finally between lecturers and reviewers when lecturers take the role of reviewers who up- and downvote and comment on reviews.

The contributions of this article are twofold: First, a report on the conception and implementation of a collaborative annotation environment that supports the collaborative peer review described in this article, and second, an analysis of the communication that happened during collaborative peer review in three courses pointing to the approach's effectiveness, but uncovering problems of the collaborative annotation system as well.

This article is structured as follows: Sect. 1 is this introduction. Section 2 introduces related work. In Sect. 3 the technological support enabling collaborative peer review is introduced. Section 4 introduces the courses in which collaborative peer review was used and presents the results of an analysis of the communication that took part during peer review. In Sect. 5 two means for supporting collaborative peer review are introduced. Section 6 concludes the article and gives perspectives for future work.

2 Related Work

This article describing collaborative peer review is a contribution to peer review and relates to computer-supported collaborative learning and collaborative annotation systems.

Peer Review. Topping defines peer review as an arrangement in which peers review their peers' work [15]. Among the positive effects of peer review, Topping mentions the additional time that is spent on the task as well as the reflective nature of peer review. For studies examining peer review of writing, he concludes that the produced reviews are "at least as good as teacher assessment and sometimes better" [15, p. 265]. Schriver [13] provides a possible explanation for that: Reviewing a great number of similar documents over a longer period of time "may actually erode [the reviewers'] skills by doing too much of the same kind of text evaluation all the time" [13, p. 245]. Cho and Schunn [2] corroborate that result as well: Their study observed that students who received feedback from a peer turned in higher quality writings than students who received feedback only from an expert. Furthermore, Cho and Schunn's results show that when more reviewers provide feedback on a document, the reviewee's final writings are of higher quality as if only a single reviewer provided feedback. The authors suggest that this may be explained by more reviewers decreasing the number of overlooked errors, as well as that the same review given by more than one reviewer is seen as more important by reviewees. In his overview of peer review in higher education, Ashenafi [1] found studies which observed that students who give good feedback are good in incorporating received feedback in their writings and one study that reported a correlation between review quality and reviewer's writing quality. Furthermore, a number of studies reported that training students in peer review improved the reviews [1].

Howard et al. [6] observed that when peer review was used in an anonymous setting, reviews were larger in length as well as more critical which suggests that anonymity supports students in being more critical of their peers' work.

One way to improve peer review may be the use of rubrics. A Rubric is "a scoring guide used to evaluate the quality of students' constructed responses" [12, p. 72], the effect of which was examined in a survey by Jonsson and Svingby [7]. As of the use of rubrics for peer review, Jonsson and Svingby found a number of studies that back the claim that rubrics have beneficial effects on the quality of the reviews as well as on students' learning.

Computer-Supported Collaborative Learning. Computer-supported collaborative learning (CSCL) studies "how people can learn together with the help of computers" [14, p. 1] According to Stahl et al., CSCL should not try to replicate what can be done physically but rather leverage technology to achieve learning scenarios that are not possible without technology. In CSCL, the majority of learning takes place through the students' interactions with each other [14].

Collaborative annotation systems allow users to annotate different types of media and share their annotations with other learners, which enables the collective collection and creation of knowledge CSCL strives to achieve [14]. According to Glover et al. [4], collaborative annotation exposes learners to different viewpoints and helps lecturers to identify misconceptions among students and weaknesses in the learning material. Thus, collaborative annotation creates a feedback loop where learning material can be revised in response to students' annotations [4]. Nokelainen et al. [10] observed that annotation activity had a positive effect on students' final grades, as well as a positive correlation between annotation quality and final grade. Students liked their peers' comments but disliked their peers' highlights on shared documents. A correlation between users' activity and examination results is reported by Yang et al. [18]. Furthermore, Yang et al.

observed that collaborative annotation generally had a positive effect on examination performance as well as a positive effect on students' attitude towards knowledge sharing. While Weng and Gennari [17] aim to improve collaborative writing with annotations, they identify a number of problems of collaborative annotation that exist in other contexts as well, among them "timely and appropriate notifications" [17, p. 579] and "comments for a wide variety of purposes" [17, p. 579].

Means for reducing the cognitive load in computer-mediated communication a form of which is enabled by collaborative annotation systems, are discussed by Hiltz and Turoff [5]. Among others, they propose the use of alarms, reminders, and notifications. In a later article, Turoff [16] further describes purposes of notifications and describes them as drivers of "communication awareness".

3 Collaborative Annotation System

The collaborative annotation system used for the peer review described in this article is implemented as part of the learning and teaching platform Backstage.[1] This section focusses on features available when annotating PDF documents, as this is the format of the students' submissions.

Users can create annotations that either refer to a passage of text or an arbitrary position on the document. When creating an annotation, users have to choose a type for the annotation, which acts as a rubric providing students with hints at directions their review should or could explore. After an annotation has been created, it is immediately shared with every other user that can access the document. An example for an annotation can be seen in Fig. 1.

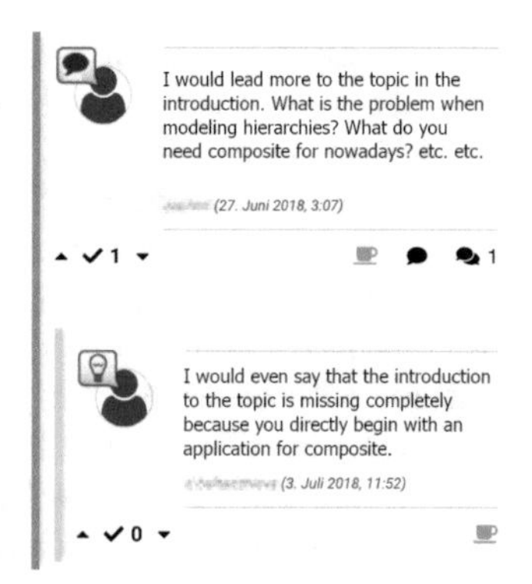

Fig. 1. Conversation between two reviewers, the second reviewer agreeing with the first reviewer and adding a further comment. The comments were translated from German and due to space reasons a comment by a lecturer was removed.

[1] https://backstage2.pms.ifi.lmu.de:8080/about.

Users can up- or downvote and comment on annotations which allows for implicit and explicit collaboration: Implicitly, by up- or downvoting which represents a form of collaborative filtering (see [8]) or explicitly, by commenting on an annotation. Both forms of collaboration prevent that the same work is done twice and aims at effects similar to those of multiple reviewers described by Cho and Schunn [2]. The annotation in Fig. 1 has been upvoted once which is indicated by the number "1" next to the checkmark and commented on once indicated by the second annotation being slightly indented following Weng and Gennari's [17] advice for threaded annotations.

4 Evaluation

The following section examines the communication that took part during peer review in three courses using the collaborative annotation system described in Sect. 3. Depending on the course, a different type of document was reviewed:

- *C1*: An article of 8 to 10 pages providing an overview of a web technology. 11 students participated in the peer review.
- *C2*: An article of 8 to 10 pages providing an overview of a software design pattern. 13 students participated in the peer review.
- *C3*: A job application of 2 to 4 pages in response to a mock job advert. 14 students participated in the peer review.

In each of the courses, a document was reviewed by two other participants. The peer review was conducted non-anonymously because students knew each others' topics in C1 and C2, and some students of C3 included their real names in their job applications.

Methods. All annotations created during peer review were retrieved from the Backstage system and classified as follows:

- Annotations that had no comments were classified as *review annotations.*
- Annotations that had at least one comment of a non-lecturer participant were classified as *conversation annotations.*

Conversation annotations were further separated based on the roles of the people involved in the conversations. For instance, an annotation created by a reviewer and commented on by the reviewee was classified as *reviewer-reviewee.* A participant commenting on their own annotation was classified as *reviewer.*

To further examine what kind of communication took part in conversation annotations, those were classified after an original classification scheme by 3 judges ($\kappa = 0.59$, moderate agreement). The communication patterns *reviewee-reviewer*, *reviewer-reviewer-reviewee*, and *reviewer-reviewer-reviewer* were omitted from the classification due to them appearing very rarely.

The communication pattern *reviewer-reviewer* was classified as follows:

- *agree*: The comment agrees with the review but does not extend upon it.
- *agree-extend*: The comment agrees with the review and extends upon it.
- *disagree*: The comment disagrees with the review but does not provide any justification for the disagreement.
- *disagree-extend*: The comment disagrees with the review and justifies the disagreement.

The communication pattern *reviewer* was classified as follows:

- *correction*: The comment corrects the original review.
- *extension*: The comment extends upon the original review.
- *clarification*: The comment clarifies the original review.

The communication pattern *reviewer-reviewee* used the same classes as *reviewer-reviewer* extended with the following classes:

- *explanation*: The comment addresses misconceptions or answers a question.
- *inquiry*: The comment inquires about the review.

Additionally, a survey was conducted at the end of each course which contained among others two parts. These parts measured the students' assessments of the effects of giving peer reviews and the received peer reviews respectively. The answers to these two parts were given on a four-point Likert scale from *strongly agree* to *strongly disagree* with no neutral option. The questionnaire raised more data than presented in this article. Due to space reasons, only the results of the aforementioned two parts of the questionnaire are presented and discussed.

Results. The number of annotations and their composition varies greatly between the courses: Though C2 had more participants, participants in C1 created more annotations (452 in C2, 664 in C1). C3 had a smaller number of annotations (219) which can be explained by the shorter length of the reviewed documents. While C1's participants created more annotations, they engaged more rarely in conversation than the participants in C2 (3.9% conversation annotations in C1, 11.5% conversation annotations in C2). Participants in C3 showed a percentage of conversations annotations similar to that of C1 (3.3%).

A breakdown of conversation annotations can be seen in Table 1: While there is no single type that is dominant across the courses, communication between reviewers and between reviewer(s) and reviewee are the most frequent forms of communication when looking at all three courses. Reviewers commenting on their own annotations made up a substantial part of conversation annotations. Overall, most conversations ended after a single comments, with conversation annotations having more than one comment (*other* in Table 1) being the minority.

Further breakdown of the three most dominant types of conversation annotations can be seen in Table 2. Only a small percentage of the conversation annotations did not extend upon the review they were an answer to (non-emphasized classes in Table 2). The majority of conversation annotations were of classes that provide further insight or require further interaction (emphasized in Table 2).

Table 1. Breakdown of conversation patterns found in conversation annotations.

Pattern	C1	C2	C3	Sum	%
reviewer-reviewee	11	16	4	31	36%
reviewer-reviewer	8	24	2	34	40%
reviewer	5	11	1	17	20%
other	2	1	0	1	4%

Table 2. Classification of the most frequent classes of conversation annotations over the courses.

Class	C1	C2	C3
reviewer-reviewer			
agree	0	5	0
disagree	2	1	1
agree-extend	4	10	0
disagree-extend	1	6	1
reviewer			
correction	0	1	0
extension	4	8	1
clarification	1	0	0
reviewer-reviewee			
agreement	4	0	1
disagree	0	1	0
agreement-extends	2	2	0
disagree-extends	2	7	1
explanation	1	6	0
inquiry	2	0	2
miscellaneous	1	4	0

Table 3 lists the results of the questionnaire. As for giving peer reviews, the participants felt that peer review gave them new ideas, improved their writing, and supported them in assessing their own performance. As of received peer reviews, the results differ between the courses: While the participants of C1 and C3 agreed with the positive effects of peer review, the participants of C2 felt that the received peer reviews were not necessarily useful for improving their work and were more inclined to agree with the statement that the received peer reviews had little to no use to them.

Discussion. The results do not exhibit a consistent picture across the examined courses: Numbers of annotations, amounts of conversation annotations, as well as their breakdown vary greatly. A difference was to be expected between C1/C2

Table 3. Results to the questionnaire for each of the courses. Strongly agree was assigned the value 1, strongly disagree the value 0. Items marked with * were worded negated, their scores subsequently inverted to receive the unnegated score. Items were shortened due to space reasons.

	C1 ($n = 8$)		C2 ($n = 6$)		C3 ($n = 4$)	
	Avg.	Med.	Avg.	Med.	Avg.	Med.
Giving Peer Review						
New ideas to improve one's own work	2.0	2.0	2.2	2.0	2.0	2.0
Better understanding of standard of work (*)	1.4	1.0	1.8	2.0	1.8	2.0
Beneficial to the learning of writing	2.1	2.0	2.0	2.0	1.5	1.5
Assessment of one's own performance	1.8	2.0	2.2	2.0	2.0	2.0
Few to none positive aspects	3.4	4.0	2.7	3.0	3.8	4.0
Received Peer Reviews						
Helped me improve my work	1.5	1.5	2.5	2.5	2.0	2.0
Beneficial to the learning of writing (*)	1.4	1.0	2.2	2.0	1.5	1.5
Opened up new writing perspectives	2.3	2.0	2.5	2.5	1.5	1.5
Little to no use for me	3.5	4.0	3.0	3.0	3.5	3.5
More valuable than lecturers' feedback	2.4	2.0	2.8	3.0	2.5	2.5

and C3 due to the differences in type and length of the documents but not between C1 and C2. Interesting insights could nonetheless be found in the data: Across all conversation patterns, the majority of the conversations consisted of a single comment. The existence of such conversations is in itself nothing unusual as there are issues that are resolved in a single comment, but it is unlikely that most issues can be resolved through a single comment.

Among the *reviewer-reviewee* conversations, neither the *inquiry* conversations nor the majority of the conversation that disagreed with the review received a further comment. While not all disagreements require further comments, the majority of inquiries requires an answer, as their existence indicates that something in the review is unclear and should be clarified by the reviewer. For *reviewer-reviewer* conversations, there are a number of disagreements as well that could potentially require resolution.

Another interesting point is that the in the vast majority of conversations, comments extended upon the content of the review providing further insight or explaining why the commenter disagreed with the review. Those comments act as feedback to the author of the annotation enabling bidirectional feedback in an arrangement where feedback traditionally only flows one-way. Another peer review environment that supports bidirectional feedback is the SWoRD system by Cho and Schunn [2], where reviewees can give one-time feedback on the reviews they received at a certain point in the peer review process.

In the scenario *reviewer*, comments were used for clarification, correction, or extension which most likely resulted from the circumstance that the annotation system provided no means for editing annotations after they have been created.

The results of the survey are consistent with previous results on peer review: Students felt that giving peer review helped them assess their own performance and gave them new ideas for their own work. Both giving peer review as well as the received peer reviews helped students with their writing skills. Students of C2 were divided whether peer review helped them to improve their work, which most likely can be explained by the quality of the received peer reviews. To ensure that every student receives an appropriate peer review, more experienced students could be assigned seminar papers of less experienced students.

5 Implications for Collaborative Peer Review

As discussed in Sect. 4, there are conversations which undeniably warrant further discussion as well as conversations which are most likely not resolved which warrant further discussion as well. Furthermore, the bidirectional feedback enabled by the conversations can only work if participants actually realize that something happened on the reviews they created or the conversations they participated in. Staying on top of a large number of conversations exerts a extraneous cognitive load on users (see [11]). In order to support users in keeping an overview of what happened and where it happened, two mechanisms for reducing the cognitive load have been adapted.

When entering the course on Backstage, alerts about interactions are displayed: Reviewees are notified if reviewers created a new annotation, and every user is notified about new messages in conversations they participated in.

After noticing that something happened, the next step is to locate where something happened. To not overwhelm the user, an "overview and detail" approach [3] for visualizing documents is utilized. In an updated version of the collaborative annotation system, the pagination of a document shows a green dot above those pages where something happened since the user's last visit. This makes it easier for users to jump to the relevant places of the document.

6 Conclusion and Perspectives

In this article, an approach using technology for blurring the phases of traditional peer review was introduced: Reviewers and reviewees alike see other reviewers' reviews immediately which allows them to engage in conversation to collaboratively work on the peer review. A side-effect of this approach is that feedback flows in both directions as opposed to traditional peer review, where feedback is only given in one direction. For an evaluation, reviews and comments of three courses were examined and classified. The majority of conversations ended after one comment but extended upon the content of the annotation, and a large number of annotations remained without comment.

The short duration of conversations and the large number of uncommented conversations may be explained by the lack of a notification mechanism: The only possibility to see what happened was to scan every page of the document for new annotations. To support future students in their collaborative peer review, two mechanisms for notification about and overview of events on the system were developed and will be evaluated in the coming terms.

Acknowledgements. The authors are thankful to Nikolai Gruschke for implementing the notification mechanisms described in Sect. 5 as part of his bachelor's thesis.

References

1. Ashenafi, M.M.: Peer-assessment in higher education-twenty-first century practices, challenges and the way forward. Assess. Eval. High. Educ. **42**(2), 226–251 (2017)
2. Cho, K., Schunn, C.D.: Scaffolded writing and rewriting in the discipline: a web-based reciprocal peer review system. Comput. Educ. **48**(3), 409–426 (2007)
3. Cockburn, A., Karlson, A., Bederson, B.B.: A review of overview+ detail, zooming, and focus+ context interfaces. ACM Comput. Surv. (CSUR) **41**(1), 2 (2009)
4. Glover, I., Hardaker, G., Xu, Z.: Collaborative annotation system environment (CASE) for online learning. Campus-Wide Inf. Syst. **21**(2), 72–80 (2004)
5. Hiltz, S.R., Turoff, M.: Structuring computer-mediated communication systems to avoid information overload. Commun. ACM **28**(7), 680–689 (1985)
6. Howard, C.D., Barrett, A.F., Frick, T.W.: Anonymity to promote peer feedback: pre-service teachers' comments in asynchronous computer-mediated communication. J. Educ. Comput. Res. **43**(1), 89–112 (2010)
7. Jonsson, A., Svingby, G.: The use of scoring rubrics: reliability, validity and educational consequences. Educ. Res. Rev. **2**(2), 130–144 (2007)
8. Linden, G., Smith, B., York, J.: Amazon.com recommendations: item-to-item collaborative filtering. IEEE Internet Comput. **1**, 76–80 (2003)
9. Nicol, D., Thomson, A., Breslin, C.: Rethinking feedback practices in higher education: a peer review perspective. Assess. Eval. High. Educ. **39**(1), 102–122 (2014)
10. Nokelainen, P., Miettinen, M., Kurhila, J., Floréen, P., Tirri, H.: A shared document-based annotation tool to support learner-centred collaborative learning. Br. J. Educ. Technol. **36**(5), 757–770 (2005)
11. Paas, F., Renkl, A., Sweller, J.: Cognitive load theory and instructional design: recent developments. Educ. Psychol. **38**(1), 1–4 (2003)
12. Popham, W.J.: What's wrong-and what's right-with rubrics. Educ. Leadersh. **55**, 72–75 (1997)
13. Schriver, K.A.: Evaluating text quality: the continuum from text-focused to reader-focused methods. IEEE Trans. Prof. Commun. **32**(4), 238–255 (1989)
14. Stahl, G., Koschmann, T.D., Suthers, D.D.: Computer-supported collaborative learning (2006)
15. Topping, K.: Peer assessment between students in colleges and universities. Rev. Educ. Res. **68**(3), 249–276 (1998)
16. Turoff, M.: Computer-mediated communication requirements for group support. J. Organ. Comput. Electron. Commer. **1**(1), 85–113 (1991)
17. Weng, C., Gennari, J.H.: Asynchronous collaborative writing through annotations. In: Proceedings of the 2004 ACM Conference on Computer Supported Cooperative Work, pp. 578–581. ACM (2004)
18. Yang, S.J., Zhang, J., Su, A.Y., Tsai, J.J.: A collaborative multimedia annotation tool for enhancing knowledge sharing in CSCL. Interact. Learn. Environ. **19**(1), 45–62 (2011)

A Moderate Experiential Learning Approach Applied on Data Science

Emilio Serrano(✉) and Daniel Manrique

Departamento de Inteligencia Artificial,
Universidad Politécnica de Madrid, Madrid, Spain
{emilio.serrano,daniel.manrique}@upm.es

Abstract. A moderate experiential learning is proposed as a framework to provide learners with significant experiences in data science. In this approach, the student learns through reflection on doing, abstract conceptualization, gamification and learning transferring; instead of being a recipient of already made content. Data science pedagogy has repeated a number of patterns that can be detrimental to the student. The proposed moderate experiential learning has been adopted together with other two learning approaches in a data science master subject for comparative purposes: a traditional learning approach, and a strict experiential learning adoption. Two evaluation studies have been conducted to compare these three different learning approaches. The results indicate that students do not actively support the strict experiential learning, but the moderate approach, where some guidelines are provided to face the realistic experience.

Keywords: Active learning · Experiential education · Project-based learning · Game-based learning · Data science · Graduate students

1 Introduction

Experiential learning (EL) is more than just getting learners to do something: *unless experiences outside the classroom are brought into the classroom and integrated with the goals and objectives of the discipline theory, students will continue to have amazing outside experiences but will not readily connect them to their in-class learning* [6]. Even when the term experiential learning is sometimes used to define any training that is interactive, with minimal lecture and slides [10], the students reflecting on their product is a fundamental part of EL. Without a careful curriculum involving structured, reflective skill building, students may never learn what we hope outside the four walls of the classroom [6]. As the Association for Experiential Education [1] claims, to ensure that EL is effective, the learner has to be actively engaged in posing questions, investigating, experimenting, being curious, solving problems, assuming responsibility, being creative, and constructing meaning. The educator and learner may experience

R. Gennari et al. (Eds.): MIS4TEL 2019, AISC 1007, pp. 11–18, 2020.
https://doi.org/10.1007/978-3-030-23990-9_2

success, failure, adventure, risk-taking and uncertainty because the outcomes of experience cannot totally be predicted. Therefore, EL is an approach that encourages collective and critical reflection, as well as individual learning [7].

Data science (DS) is an interdisciplinary field devoted to extracting knowledge from complex big data [5]. The great diversity of applications and the growing demand of experts in the DS field has made courses, books and manuals in DS proliferate. The standard pedagogical method in DS that can be appreciated basically consists of four steps: the explanation of different DS techniques; the detail on some specific approaches or paradigms; the illustration of these paradigms using toy, and well known, datasets [11]; and the assignments with a straightforward application of the ideas previously exposed using some DS framework [4]. We also adopted this traditional learning approach over the last few years in a data science related course as part of a University master degree program. This experience revealed some limitations of this method. Firstly, students showed difficulties in selecting relevant information about how different machine learning approaches work and what kind of problems they are suitable for. As a result, the student usually obviated the details and data of a concrete problem. As Witten et al. declare in [11], nothing replaces a good understanding of the data. Finally, the creativity in solving problems is considerably restricted because DS is perceived as the application of well-known solutions to well-known problems.

EL would naturally mitigate these tendencies when learning DS because it focuses on problems to be solved instead of on specific methods. In addition, starting with realistic experiences gives students more experience in real-world problems. More importantly, creativity and divergent thinking are encouraged when searching for solutions to a concrete experience. The existence of different dataset repositories on which to build knowledge offers a privileged breeding ground for designing a DS course as a series of experiences in real-world problems [2,9].

This paper presents a moderate experiential learning approach, supported by open and free software framework [8]. This approach has been deployed in a deep learning course, which is part of the official Master's Programme in ICT Innovation: Data Science (EIT Digital Master School). Deep learning is nowadays a very demanding working and studying area due to its dramatically improved results in many domains such as classification, computer vision and sequential data problems. These are, precisely, the three units in which the course is divided. For comparison purposes, the traditional learning approach is involved in the first unit: deep artificial neural networks for classification problems. The proposed approach in the second unit: computer vision; and a strict experiential learning adoption is deployed in the third unit for natural language processing, i.e. a sequential data problem. The experience gathered over the 2016–17 academic year is also presented in terms of the level of students' satisfaction for each of the three deep learning course units/learning approaches.

2 The Moderate Experiential Learning Approach

EL theory is typically represented by a four-stage learning cycle [3]: effective learning involves progressing through this cycle: first, having a concrete experience or situation; then, the observation of and reflection on that experience; later, the formation of abstract concepts (analysis) and generalizations (conclusions); and finally, testing them by active experimentation, resulting in new experiences (iterations in the cycle). In the scope of our moderate EL approach, this learning cycle is reviewed and instantiated for the computer vision unit, within the deep learning course.

Deep learning is a second year 3-ECTS (European credit transfer and accumulation system) master course, with a duration of one semester (15 weeks) and 81 h of student workload distributed as follows: 30 h to fifteen 2-h face-to-face classroom sessions, one per week, and the remaining 51 h to workgroup and individual activities. The course kicks off with a one-day face-to-face session where the learners have the chance to meet the instructor or facilitator. Since this is a second-year course, learners already meet each other, what would be another objective of this session in case of a first-year course. The instructor introduces the course topics, presents the learning objectives, and discusses the most significant knowledge to acquire. Then, the instructor explains the three different learning methods that will be adopted for each of the three units: traditional learning for deep artificial neural networks for classification problems, the proposed moderate EL for computer vision and EL for natural language processing. The instructor also encourages the students to form workgroups of three or four members during the rest of this course week.

Every week, there is a two-hour face-to-face session where some experiential learning activities take place depending on the learning approach for the unit. Additionally, the instructor presents the most important contents to learn over the following week. Students have also the opportunity to put in common questions to be discussed. Learners can meet the instructor, individually or in group, six hours a week to clarify contents and receive support on how to solve problems or experimental activities. Students regularly meet at their discretion in workgroups to discuss the experiences presented in class or possible solutions, to accomplish the active experimentation or to solve the assessment activities. An assessment activity is set immediately after each unit has finished, related to a real-world problem that involves a dataset extracted from a public web repository. Learner evaluation also considers the scores achieved in the solutions given to the assessment activities, which are the same for all the members of a workgroup.

2.1 The Three Learning Approaches

The first unit, deep artificial neural networks for classification problems, takes five weeks. It follows the classical flow in a DS course. Firstly, artificial neural networks theory is explained. Secondly, practical advice in solving problems are described along with deep learning frameworks. Finally, an assessment activity to solve in workgroup a classification problem using a public dataset is set to apply the explained ideas.

The second unit, computer vision, takes the following five weeks. It adopts a moderate EL approach. After introducing the topic, a contest is proposed for the workgroups to use the methods learned in the previous unit to a computer vision problem using a specific dataset. The experience allows a reflective observation of the low accuracy achieved, and an abstract conceptualization of some of the challenges of computer vision. Then, convolutional neural networks are explained. This allows students to retake the contest and observe the improvement achieved by the new ideas introduced in the course. Finally, the same approach is followed for a transfer learning problem, proposing a new experience for images object classification as an assessment activity. The key in this moderate experiential approach adopted in this unit is that students get their prior knowledge challenged by new problems. Learners have time to try known methods to new situations and to reflect on the results. Moreover, the contests act as a game-based approach for the EL [9]. Students are not required to research new methods for the new experiences proposed as planned in a stricter EL approach.

The third unit, natural language processing, adopts a strict experiential learning approach to learn the educational contents for four weeks. The learners, also in workgroups, face a realistic experience: they are asked to design a solution to predict the relevance of an article headline regarding its body from a dataset. First, workgroups carry out an investigation into the problem and present a manuscript with the solution design. Bibliographic references discussing natural language processing methods are provided afterwards to reflect on possible changes in the solution. Then, the workgroups have the opportunity to present their solutions to be discussed with the other groups in a face-to-face session moderated by the facilitator. Finally, a brief lecture on natural language processing concepts and how to apply them to the case study is offered. The assessment activity consists of delivering a report with the final solution design. A key differentiator of this strict EL adoption is that it makes the learners feel free in the investigation and proposal of solutions to a problem, instead of following instructor instructions.

3 Results

Two evaluation studies have been conducted to analyze the level of learners' satisfaction using three different learning approaches for the deep learning course. The first study evaluates each learning approach with the same set of questions. In the second study, each learning approach is voted against the other two based on some affirmations. This course has been taught over the 2016-17 academic year to computer science graduates as part of a Data Science Master Degree Program. A total of 25 students attended the course, from which 24 participated in the study. They come from European countries and are of very similar ages, ranging from 22 to 24 years old. The course was taught by the same teachers using the three learning approaches. Additionally, each learner is exposed to the three learning methodologies, avoiding this way the residual variation due to differences between subjects.

Table 1. Descriptive statistics for learners' satisfaction, evaluated by the three questions Q1, Q2, and Q3 for each of the three learning methodologies adopted

Q:U	*N*	Mean	Std. Dev.	Std. Error	Min.	Max.	95% CIM	
							Lower B.	Upper B.
Q1:U1	24	3.38	1.06	0.2164	1	5	2.95	3.80
Q1:U2	24	4.50	0.59	0.1204	3	5	4.26	4.74
Q1:U3	24	4.13	0.68	0.1387	3	5	3.85	4.40
Q2:U1	24	3.58	0.78	0.1592	2	5	3.27	3.89
Q2:U2	24	4.46	0.51	0.1041	4	5	4.25	4.66
Q2:U3	24	4.08	0.97	0.198	2	5	3.69	4.47
Q3:U1	24	3.08	0.97	0.198	1	5	2.69	3.47
Q3:U2	24	4.00	0.88	0.1806	2	5	3.65	4.35
Q3:U3	24	4.21	0.93	0.191	2	5	3.84	4.58

Data were obtained from a questionnaire administered to students at the end of the course. The questionnaire is divided into two sections: one for each of the two evaluation studies. The first section comprises the same three questions for each learning methodology. The second section comprises four affirmations in which the learner has to decide (or vote), for each of them, the course unit/learning methodology that best fits it.

3.1 Learning Approaches Evaluation

The first section of the questionnaire administered to the students to evaluate each of the three learning approaches comprises the following three questions: Q1: *the unit content meets my training needs*, Q2: *what I learned will be applicable in my job*", and Q3: *the applied methodology, technical resources and teaching materials were appropriate*. Participants responded to these questions based on a five-point Likert scale, ranging from strongly disagree (scored as 1) to strongly agree (scored as 5). The learning methodology served as the independent variable, with three levels: traditional learning for unit 1 (U1), moderate EL for unit 2 (U2) and strict EL for unit 3 (U3). The dependent variable for this study is the level of learners' satisfaction, represented by the learners' responses to Q1, Q2, and Q3. Table 1 shows descriptive statistics for the responses to each question related to each unit. From left to right: number of responses gathered, standard deviation, standard error, minimum and maximum values, and 95% confidence interval with lower and upper bounds.

Since the same students faced the three learning methodologies and, therefore, the same subjects responded to the questionnaire, a one-way repeated-measures analysis of variance (rANOVA) was conducted to gather empirical evidence of whether the differences between the means are statistically significant. Therefore, the learning methodology (the independent variable) is the within-subjects factor. The Shapiro-Wilk and Levene tests confirm that the rANOVA

Table 2. Tukey HSD test for Q1. $p < 0.05$

Learning condition a	Learning condition b	Mean difference	Sig.
Q1:U1	Q1:U2	−1.125*	.000*
Q1:U1	Q1:U3	−0.75*	.005*
Q1:U2	Q1:U3	0.375	.055

assumptions (normality and homoscedasticity, respectively) are met. Since there are three groups with 24 observations each, the degree of freedom, *df*, between groups is 2, and within groups is 69. The results of the three rANOVA tests show that the null hypothesis (responses to questions Q1, Q2, and Q3 are the same for each of the three learning conditions) are rejected since $F(df = 2/69) = 12.28$, $p < 0.01$; $F(df = 2/69) = 7.67$, $p < 0.01$, and $F(df = 2/69) = 9.92$, $p < 0.01$, respectively. Thus, the learners' satisfaction depends on the learning approach.

As there are sizeable differences between the three groups of each question Q1, Q2, and Q3 concerning the learning approach, the Tukey HSD (honestly-significant-difference) test was used to make post hoc comparisons to demonstrate where are the statistically significant differences. Table 2 shows the results of these multiple comparisons for Q1, which are similar to Q2, and Q3, considering that the significance level for the mean difference is $p < 0.05$, marked by an asterisk. The conclusions that arise from Table 2 are the same for Q2 and Q3: there are significant differences between the classical learning approach (U1) and the moderate EL (U2), and between the classical approach and the strict EL (U3). However, there are no differences between the two EL approaches (U2 and U3). From these results, the homogeneous subsets for $\alpha = 0.05$ have been also statistically calculated, as a function of the mean, to group the three learning conditions applied. The result was the same: a subset formed exclusively by the classical learning approach and another subset containing the proposed moderate EL, plus strict EL. The values obtained from Q1, Q2, and Q3 for either moderate or strict EL are, statistically, greater than for learners responding to the questionnaire in relation to the classical learning approach. It is also noteworthy that, although the means for Q1, Q2, and Q3 using the proposed moderate EL approach appears to be higher (4.50, 4.46, and 4.00, respectively. See Table 1) than using the strict EL approach (4.13, 4.08, and 4.21, respectively), there are no significant differences in terms of the learners' satisfaction.

3.2 Confronting the Three Learning Approaches

The second section of the questionnaire administered to students at the end of the course is involved in this second evaluation study. Four affirmations are presented: A1: *my favourite methodology is the one used in...*, A2: *the methodology that allows me to learn the most is...*, A3: *the methodology that (I believe) allows the longest lasting learning is...*, and A4: *the methodology that is closest to a data scientist's daily tasks is...*. The learner is asked to vote the learning condition that best fits each of these affirmations. The aim of this second evaluation study

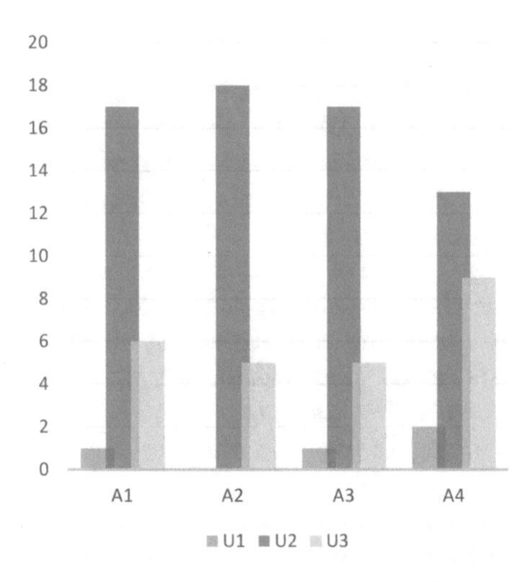

Fig. 1. Votes for each learning condition

is to deepen the comparison of the three learning approaches, especially in the cases of moderate and strict EL, since both were grouped in the same homogeneous mean subset in the previous evaluation study. Figure 1 shows a histogram of the votes cast for each learning condition. The affirmations are represented in the abscissa axis and the votes cast for each learning condition in the ordinate axis. It is clear that the preferred learning condition is the moderate EL approach (represented by blue bars), except for the affirmation marked as A4. In this case, moderate and strict EL approaches are closed to each other, with 13 and 9 votes, respectively. Therefore, it can be concluded that although these two EL approaches seem to be equivalent from the first evaluation study, students prefer the proposed moderate approach when they have to decide between the moderate against the strict approach.

4 Conclusions

This article presents what is classed as a moderate experiential learning approach, since students get their prior knowledge challenged by new problems, instead of freely researching new methods for the new experiences proposed. It is a general-purpose learning approach, supported by an open and free software framework, that can be applied when realistic experiences are available. This is the case of data science, where the proposed moderate EL approach is now being successfully applied to computer science graduates. We have also presented the results in terms of learners' satisfaction from experience gathered over the 2016–17 academic year, comparing three different learning approaches for a deep learning course: the classical flow followed in data science courses, the proposed moderate EL, and a strict EL adoption. Two evaluation studies have been conducted, using data obtained from a questionnaire administered to students at the end of the course.

The first study involved one-way repeated-measures analysis of variance. The results provide empirical evidence that the level of learners' satisfaction is higher in the case of the moderate and strict EL than for the classical approach. However, there are not statistically significant differences between moderate and strict EL methods, even though the learners' responses to the questions evaluated in this study favor, on average, the moderate EL prescriptions. The second study consisted of voting for each of the three learning approaches under study to resolve the statistical tie between the moderate and strict EL approaches. This way the student had to decide the most satisfying learning condition against the other two. The results achieved show a superiority of the proposed moderate EL approach, with more than 70% of the votes cast in three out of four affirmations evaluated, and also winning the vote in the last case, although with a closer result.

Future research work focuses on to strengthen the findings and conclusions arisen from this study by increasing the sample sizes and a more in-depth analysis of where the differences are between the moderate and strict EL.

Acknowledgements. This work was supported by Universidad Politécnica de Madrid under Grant IE1718.1003; Ministerio de Economía, Industria y Competitividad, Gobierno de España (ES) under Grant TIN2016-78011-C4-4-R, AEI/FEDER, UE.

References

1. Association for Experiential Education: The principles of experiential education in practice (2018). http://www.aee.org/index.php?option=com_content&view=article&id=110:what-is-ee&catid=20:other&Itemid=260
2. Bennett, J., Lanning, S.: The Netflix prize. In: KDD Cup and Workshop in conjunction with KDD. ACM, San Jose (2007)
3. Kolb, D.A.: Experiential Learning: Experience as the Source of Learning and Development, 2nd edn. Pearson Education, Upper Saddle River (2015)
4. Kuhn, M.: Caret: classification and regression training (2018). https://CRAN.R-project.org/package=caret
5. Mayer-Schonberger, V., Cukier, K.: Big Data: A Revolution That Will Transform How We Live, Work, and Think. Houghton Mifflin Harcourt, Boston (2013)
6. Qualters, D.M.: Making the most of learning outside the classroom. New Dir. Teach. Learn. **124**, 95–99 (2010)
7. Reynolds, M., Vince, R.: Handbook of Experiential Learning and Management Education. Oxford University Press, Oxford (2008)
8. Serrrano, E., Manrique, D., Amador, E.: JupyterDS: a software tool for experiential learning in data science. Registration in the intellectual property registry of Madrid No. M-004029/2018 (2018)
9. Shiralkar, S.: IT Through Experiential Learning: Learn, Deploy and Adopt IT Through Gamification. Apress, New York (2016)
10. Silverman, M.L.: The Handbook of Experiential Learning. Pfeiffer, San Francisco (2007)
11. Witten, I.H., Frank, E., Hall, M.A.: Data Mining: Practical Machine Learning Tools and Techniques. Morgan Kaufmann, Burlington (2011)

The Automated Grading of R Code Snippets: Preliminary Results in a Course of Health Informatics

Anna Maria Angelone and Pierpaolo Vittorini(✉)

Department of Life, Health and Environmental Sciences, University of L'Aquila, 67100 L'Aquila, Italy
pierpaolo.vittorini@univaq.it

Abstract. Automated grading tools either execute the submitted code against sample data or statically analyse it. The paper presents a tool for the automated grading of R code snippets, submitted by students learning Health Informatics in the degree course of Medicine and Surgery of the University of L'Aquila (Italy). The tool performs a static analysis of the R commands, with the respective output, as well as of the sentences written in natural language. The paper details the problem in general and through examples. Then, it describes the proposed solution and reports on the comparison between the automated grading and the human one. Finally, the paper ends by describing the foreseen use of the tool and the needed improvements.

Keywords: Summative assessment · Formative assessment · Automated grading

1 Introduction

Individual assessment is a valuable tool in this society [12]. In the context of learning, assessment can essentially be divided into formative and summative assessment [10]. Formative assessment takes place during a course, as a means of checking on student learning progress, by the teacher or by the students themselves. Summative assessment takes place at the end of a period of study to determine examination outcomes. Within any assessment method, question types may vary. They can be short essay type questions, true or false type questions or multiple-choice questions. In computer science courses, the specific type of code-snippets questions is also common.

Human grading of code-snippets questions is a tedious and error-prone task, a problem particularly pressing when such an assessment involves numerous students. One possible solution to this problem is to automate the grading process so that it can facilitate teachers in the correction and enable students to receive immediate feedback. Many tools implementing an automated grading of code-snippets already exist [15]. The capabilities of such systems range from the compilation and execution of students' programs against test data, up to the static analysis of the students' source code. They are used in both formative and summative assessment, as well as to encourage students to improve their assignments before the final submission.

Given these premises, the paper presents a tool for the automated grading of R code snippets, submitted by students learning Health Informatics in the degree course of Medicine and Surgery of the University of L'Aquila (Italy).

R. Gennari et al. (Eds.): MIS4TEL 2019, AISC 1007, pp. 19–27, 2020.
https://doi.org/10.1007/978-3-030-23990-9_3

The application scenarios for both formative and summative assessment are the following. During the course, the students perform exercises that require the execution of R commands and the interpretation of the results. As a formative assessment instrument, the developed tool provides to students both a feedback and an estimated evaluation of the quality of the submitted solution. At the end of the course, the students perform a test, similar to the exercises faced during the course, without the help of the tool. Nevertheless, as a summative assessment instrument, the tool is used by the teacher to support the manual correction activities. Accordingly, the foreseen educational benefits are twofold. For students, we aim to support their understanding of the commands, the interpretation of the results, and – as a consequence – increase the final exam outcome. For teachers, to reduce their workload, either in terms of manual correction times, but also in a reduced number of the returning students.

The paper is organised as follows. It first details the problem in general and through examples. It then describes the proposed solution, reports on the comparison between the automated grading and the human one, and – based on the results – ends by describing the possible use of the tool and the needed improvements.

2 Problem Description

Students learning health informatics need to master both general and specific computer science skills. As for the latter, the knowledge of how to execute statistical analyses of biomedical data is fundamental, since statistical analyses are present in scientific papers and are mandatory for adding scientific evidence to their activities (e.g., thesis).

The statistical analysis of biomedical data is a rather vast subject. It includes both descriptive and inferential statistics, some of them being very specific. Accordingly, the subject of Health Informatics of the degree course of Medicine and Surgery of the University of L'Aquila includes a specific topic on how to perform statistical analyses of biomedical data

Let us consider the following dataset:

Subject	Surgery	Visibility	Days
1	A	7	7
2	A	5	7
	...		
10	B	16	12
	...		
20	C	19	4

The data regards 20 subjects (variable Subject) that underwent three different surgical operations (variable Surgery). We observe the scar visibility (variable Visibility) in terms of ranks ranging from 1 (the best) to 20 (the worst). We also measure the hospital stay (variable Day).
You are required to:

1. calculate the mean (with confidence intervals) and the standard deviation of the hospital stay;
2. calculate the absolute and relative frequencies (with confidence intervals) of the surgical operations;
3. verify if the hospital stay can be considered as extracted from a normal distribution;
4. comment on the result;
5. calculate the median, the 25th and 75th percentile of the hospital stay for the different surgical operations;
6. verify if the aspect of the scar is different within the different surgical operations;
7. comment on the result.

Submit as solution a text containing the list of R commands with the respective output, as well as your interpretation of the analyses 3 and 6.

Fig. 1. A sample exam

using R [14]. To pass the final exam, students are required to execute statistical analyses in R and to correctly explain the achieved results.

Figure 1 shows the text of a sample exam. To solve the assignment, the student has to reflect on different points, like the variables' type (if quantitative or qualitative), the number of categories, what a p-value means and what it suggests. By referring to the sample exam, let us take into account the sixth point of the assignment. Since the scar visibility is qualitative and the number of different surgeries is three, the student should use a Kruskal-Wallis test. Such a test is executed in R through the command:

```
kruskal.test(data = ex11, Visibility ~ Surgery)
```

which would return the following output:

```
	Kruskal-Wallis rank sum test

data:  Visibility by Surgery
Kruskal-Wallis chi-squared = 9.8959, df = 2, p-value = 0.007098
```

By looking at the p-value, which is less than 0.05, the student should then conclude that the scar visibility is different with respect to the different types of surgical operations. This conclusion is the solution for point 7 of the assignment.

Therefore, the solution of an assignment is a text containing R commands, their output and sentences written in natural language.

Given these premises, the problem is to assign an accurate grade to solutions given to such a kind of assignments, by taking into account that they contain commands, outputs and sentences. The next sections overview the solution, formalise it, describe its implementation and show the results of the comparison of the automated grading vs the human one.

3 The Adopted Solution

As previously described, the problem we face is twofold, i.e., the automated grading of the R code plus the understanding of the comments. Both tasks have been independently studied in the scientific literature. As for the automated grading of programming assignments [15], the existing tools can belong to different categories, depending on the approach, characteristics and strategy. The approach can be either instructor-centred, student-centred or hybrid, if the tool – respectively – either support the instructor, the students or both. As for the latter approach, hybrid tools may implement preliminary validation (e.g., [4]), partial feedback (e.g., [7]) or enable instructor review (e.g., [8]). In terms of characteristics, the main features of the existing tools regard the electronic submission, automated checking and grading, as well as providing immediate feedback (e.g., [11]). Finally, as for the correction strategy, they either execute the submitted code against sample data or statically analyse the source code [15]. Regarding the short answer grading problem, there have been many publications, especially in the last decade. An extensive analysis of such a literature can be found in [3].

As for the scope of the paper, it is worth summarising that – usually – the existing tools either compute a real-valued score (e.g., from 0 to 1) or to assign a label (e.g., correct or irrelevant) to a student response (e.g., [16]).

Accordingly, the adopted solution offers an hybrid approach, with characteristics ranging from the electronic submission of assignments, feedback to students, instructor review and automated grading. To the best of our knowledge, this is the first proposal specifically focusing on R, with mixed assignments containing both code and comments written in natural language. Subsection 3.1 discusses on the technical solution concerning the automated grading, whereas Subsect. 3.2 shows the main characteristics of the system.

3.1 Automated Grading

The tool works as follows. First, it analyses the correct solution (given by the teacher), the submitted one (given by the student) and returns two lists of triples containing the command, its output and the possible comment. Second, it calculates a distance between them as the weighted sum of the number of missing commands (i.e., commands that are in the correct solution, but not in the submitted one), the number of commands with a different output (i.e., commands that are in both solutions, but differ in terms of output) and the distances between the correct and the submitted comments.

Given the above, the distance between the solutions is defined as follows:

$$d = w_m \cdot \ M + w_d \cdot \ D + \sum_i w_c \cdot \ (1 - C_i) \tag{1}$$

where w_m is the weight assigned to the missing commands, M is the number of missing commands, w_d is the weight assigned to the commands with different output, D is the number of commands with different output, w_c is the weight assigned to the distance between the comments, and C_i is the distance between the i-th comments, if present. For the experiments reported in the paper, we used $w_m = 1, w_d = 0.1, w_c = 0.1$.

As for the analysis of the text, so far the tool calculates the distance between the comments using the Levenshtein string similarity distance [13] divided by the length of the longest string[1]. It is clear that string similarity distances measures the lexical rather than semantic similarity [9] and therefore they can score as different, strings conveying the same concept (and viceversa). For instance, given the sentence "The difference is statistically significant", the adopted string similarity distance would consider the sentence "The difference is not statistically significant" (that conveys the opposite concept) only 0.085 distant. A supplementary discussion about this point is in Sect. 5.

As for the final step, the tool converts such a distance into the final grade. In Italy, a grade is a number ranging from 0 to 30, plus a possible additional mention of "cum laude", customarily considered as 31. An exam is passed with a grade higher or equal to 18. Accordingly, with a simple proportion, the tool

[1] This division is adopted so to return a distance in the range [0, 1].

converts a distance equal to zero to the grade of 31, and a maximum distance (that depends on the number of commands and comments needed to solve the assignment) as the grade of zero.

3.2 Implementation

The section shows how the tool is currently used within the UTS (Univaq Testing Suite) system [2] by both the students (i.e., formative assessment) and the professors (i.e., summative assessment).

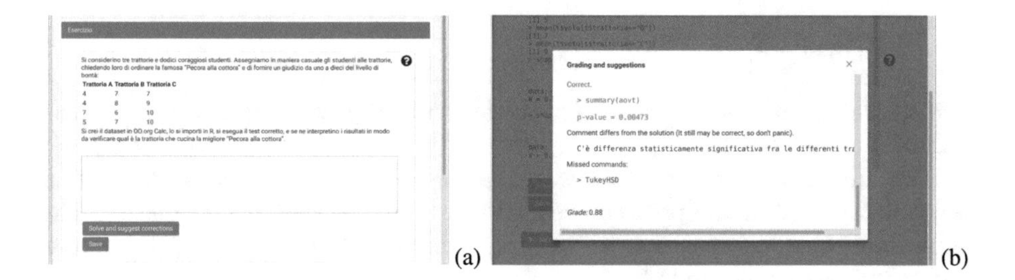

Fig. 2. Formative assessment (assignment in Italian language)

Figure 2 refers to the interfaces proposed to a student while trying to solve an exercise: on the left the interface before starting the assignment, on the right the dialog opened when requesting the automated feedback and grading. As shown, before starting to solve the exercise (Fig. 2-a), the student can see the text of the assignment (on the top), can write the solution in a text area (on the centre) and has two buttons (on the bottom) to request the automated feedback and save the solution. If the student requests the automated feedback and grading, the system opens a dialog which contains the commands/outputs/comments that are correct, missing, differ from the provided solution, as well as the calculated grade.

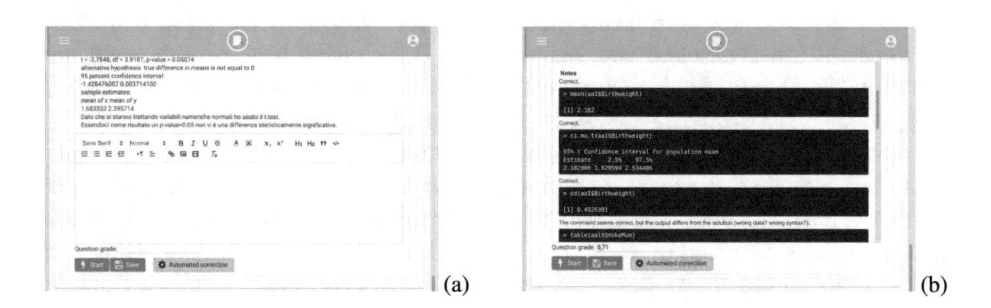

Fig. 3. Summative assessment (assignment in Italian language)

Figure 3 refers to the interfaces proposed to a student while trying to solve an exercise: on the left the interface before starting the correction, on the right the interface after the professor click on the "Automated correction" button. As shown, before starting the correction (Fig. 3-a), the professor can see the text of the assignment (on the top), has a text area where to write the correction notes (on the centre), another where to enter the grade (right below the previous one) and two buttons (on the bottom) to save the notes and to start the automated correction. If the professor starts the automated correction, the tool is activated (Fig. 3-a): the text areas are filled with the results of the analysis and with the automated grade.

4 Evaluation

The section reports on the quality of the automated grading with respect to the manual one as follows.

The automated grading tool was used to grade the solutions of a set of assignments used in the previous year of the course of Health Informatics. The results were then compared with the grades already assigned during the exams. The comparison consisted of:

- in terms of the numerical grade:
  - the intraclass correlation coefficient [1], in order to measure the level of agreement of the automated/manual grading;
  - the linear regression [17], so to understand if and how the manual grade can be "predicted" by the automated one;
- in terms of a dichotomic fail/pass grade:
  - the Cohen's Kappa [6], so to measure the level of agreement of the automated/manual grading.

The results follow.

In total, we analysed a total of 69 solutions. The solutions belonged to 11 different assignments, all with the same structure of the assignment in Fig. 1. The solutions were submitted during 8 different assessment rounds. The average grade was 27/30 (s.d. 5), and the 10% of solutions was considered unsatisfactory to pass the exam. By comparing the manual and automated grades, the intraclass correlation coefficient was measured as 0.784, with 95% confidence intervals of $[0.674, 0.860]$. Such a value can be interpreted as an excellent indication of agreement between the two sets of grades [5]. Figure 4 shows all the manual/automated grades and the linear regression line with confidence intervals, where the automated grade is the independent variable and the manual grade the dependent variable. The linear regression model was statistically significant ($p < 0.000$) and resulted in an $R^2 = 0.740$. This result indicates that the data are almost close to the fitted line, and therefore there is a good linear relationship between what is measured automatically and what was assessed by the professor. By looking at the regression line (MANUAL $= 5.41 + 0.87 \cdot$ AUTOMATED), it becomes clear that the automated grading tool is more "conservative" than

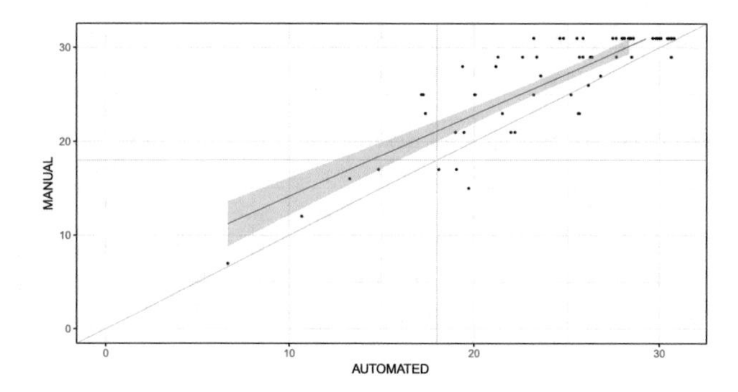

Fig. 4. Linear regression

the human evaluator. By transforming the numerical grades into dichotomic pass/fail outcomes, the agreement measured with the Cohen's Kappa was 0.52. Such a value is not satisfactory. The reason for this unsatisfactory results is that the automated grading tool erroneously considered three exams failed and three passed, when they should have been the opposite. These exams are the points in Fig. 4 in the higher-left and lower-right portion of the Cartesian plane. By instead ignoring all the automated grades close to the passing grade of 18, i.e., within the range of [16, 20], we instead obtain a perfect agreement.

In summary, the results suggest that the automated grading tool may provide useful support to the professor in correcting the assignments because the automated grade is adequately accurate and because the tool returns the commands, outputs and comments that were absent or that differ from the correct solution. On the other hand, the results suggest to use it more carefully with the students since it returns false positive/negatives when close to the pass/fail threshold.

5 Discussion and Future Work

Given the previous results, the future work is twofold.

From an implementation viewpoint, two activities must be carried out. The first regards the improvement of the tool, by increasing the agreement and the R^2 coefficient, so to minimise the number of false positives/negatives. The second concerns the natural language processing task. Even if the results are satisfying, essentially because the weight assigned to the similarity measure is not prominent, it is clear that an improvement in such a respect is mandatory. So far, the authors are evaluating the different approaches currently available, so to select the one that best fits our problem.

From an investigation viewpoint, as the results suggested and as discussed in the introduction, we introduced the automated grading tool in the Health Informatics course as follows:

- students: only during the formative assessment, i.e., as a tool suggesting how to improve the solutions to exercises before the final submission;

- professors: during both the formative and summative assessments, i.e., as an aid for correcting the solutions to exercises and exams.

In such a context, we aimed at measuring: (i) from the students perspective, whether the tool is usable and if it helps in improving the formative outcomes; (ii) from the professors viewpoint, if the tool actually speeds up the grading process. At the time of writing, the study is still in progress and therefore no results are reported. Nevertheless, unstructured interviews with few students that used the system indicate an overall appreciation and a foreseen usefulness of the tool.

References

1. Bartko, J.J.: The intraclass correlation coefficient as a measure of reliability. Psychol. Rep. **19**(1), 3–11 (1966)
2. Bernardi, A., Innamorati, C., Padovani, C., Romanelli, R., Saggino, A., Tommasi, M., Vittorini, P.: On the design and development of an assessment system with adaptive capabilities. In: Methodologies and Intelligent Systems for Technology Enhanced Learning, pp. 190–199. Springer, Cham (2019)
3. Burrows, S., Gurevych, I., Stein, B.: The eras and trends of automatic short answer grading. Int. J. Artif. Intell. Educ. **25**(1), 60–117 (2015)
4. Choy, M., Lam, S., Poon, C.K., Wang, F.L., Yu, Y.T., Yuen, L.: Design and implementation of an automated system for assessment of computer programming assignments. In: Advances in Web Based Learning, ICWL 2007, pp. 584–596. Springer, Heidelberg (2007)
5. Cicchetti, D.V.: Guidelines, criteria, and rules of thumb for evaluating normed and standardized assessment instruments in psychology. Psychol. Assess. **6**(4), 284–290 (1994)
6. Cohen, J.: A coefficient of agreement for nominal scales. Educ. Psychol. Measur. **20**(1), 37–46 (1960)
7. Edwards, S.H., Perez-Quinones, M.A., Edwards, S.H., Perez-Quinones, M.A.: Web-CAT: automatically grading programming assignments. In: Proceedings of the 13th Annual Conference on Innovation and Technology in Computer Science Education, ITiCSE 2008, vol. 40, p. 328. ACM Press, New York (2008)
8. Georgouli, K., Guerreiro, P.: Incorporating an automatic judge into blended learning programming activities. In: Advances in Web-Based Learning, ICWL 2010, pp. 81–90. Springer, Heidelberg (2010)
9. Gomaa, W.H., Fahmy, A.A.: A survey of text similarity approaches. Int. J. Comput. Appl. **68**(13), 13–18 (2013)
10. Harlen, W., James, M.: Assessment and learning: differences and relationships between formative and summative assessment. Assess. Educ. Princ. Policy Pract. **4**(3), 365–379 (1997)
11. Joy, M., Griffiths, N., Boyatt, R.: The boss online submission and assessment system. J. Educ. Resour. Comput. **5**(3), 2–es (2005)
12. Knight, P., Yorke, M.: Society for Research into Higher Education: Assessment Learning and Employability. Society for Research into Higher Education & Open University Press, Maidenhead (2003)
13. Levenshtein, V.I.: Binary codes capable of correcting deletions, insertions and reversals. In: Soviet Physics Doklady, vol. 10, p. 707 (1966)

14. R Core Team: R: A Language and Environment for Statistical Computing (2018)
15. Souza, D.M., Felizardo, K.R., Barbosa, E.F.: A systematic literature review of assessment tools for programming assignments. In: 2016 IEEE 29th International Conference on Software Engineering Education and Training (CSEET), pp. 147–156. IEEE, April 2016
16. Sultan, M.A., Salazar, C., Sumner, T.: Fast and easy short answer grading with high accuracy. In: Proceedings of the 2016 Conference of the North American Chapter of the Association for Computational Linguistics: Human Language Technologies, pp. 1070–1075. Association for Computational Linguistics, Stroudsburg (2016)
17. Weisberg, S.: Applied Linear Regression. Wiley, Hoboken (2013)

Learning by Fiddling: Patterns of Behaviour in Formal Language Learning

Niels Heller(✉) and François Bry

Ludwig Maximilian University of Munich, Munich, Germany
niels.heller@ifi.lmu.de

Abstract. This article reports on patterns of behaviour among students learning several of the formal languages taught in STEM using a multi-language text editor that detects syntactic errors. Conveying formal languages such as programming languages and mathematical formalisms is an essential and difficult, yet little investigated, aspect of STEM education. An intensive evaluation based on the use of the editors in teaching seven different programming languages in two university courses in computer science has first manifested a significant correlation between the editors' use and examination success. The evaluation has also unveiled interesting patterns of behaviour in the editors' use among students succeeding at the examination: They not only extensively used the editors but also used them in a manner which can be called "code fiddling", that is, experimenting with code examples from the learning material by modifying it, while using the editors over longer periods of time than non-fiddling students. The evaluation has also shown that the students consider the editors useful for their learning. This article reports on the afore-mentioned educational approach to teaching STEM formal languages and on its evaluation. It furthermore indicates implications for future research and for STEM teaching.

Keywords: Computer science education · Programming · Formal languages · Case study

1 Introduction

Learning to write programs is often reported to be of particular difficulty for beginners, which is often attributed to the requirement of simultaneously mastering syntax and semantics of a programming language [4,6]. Yet, this requirement does also apply to many other STEM fields, as these typically rely on abstract formalisms (such as algebraic expressions in mathematics of structural formulas in chemistry), the semantics and syntax of which have to be learned, explored, and finally put to use by the students.

This article reports on an evaluation of an online learning tool aimed at facilitating the learning of such formal languages. The tool consists of a web-based multi-language text editor supporting several "practical" programming

R. Gennari et al. (Eds.): MIS4TEL 2019, AISC 1007, pp. 28–36, 2020.
https://doi.org/10.1007/978-3-030-23990-9_4

languages (such as Python or Java) several theoretical programming languages (such as LOOP used for proving results in theoretical Computer Science but not for programming), and several mathematical formalisms (such as those expressing transformations of logical expressions).

Theoretical programming languages are similar to "practical" programming languages: They follow a non ambiguous syntax and can be executed (hopefully) yielding the intended result. Yet, theoretical programming languages also differ from practical programming languages: They are focused at one single concept (LOOP, for instance, is used to explore computability, that is, which functions can be expressed as fixed-length loops), and have a much simpler syntax. Mathematical formalisms also follow a non-ambiguous syntax, yet they typically express formal processes (transforming algebraic formulas, expresses the process of solving an equation for a specific variable). Therefore, mathematical "code" expressed in a formal language of mathematics is "verified", not "executed": In the case of equivalent transformations, such an evaluation would most likely include checking all transformation steps for correctness.

With the help of the multi-language editor evaluated in this article, programs as well as formal expressions can be composed and tested. The tool then either reports syntax errors found in the code or displays evaluation results.

While the tool supports several "practical" programming languages, the evaluation reported about in this article is based solely on the use of five theoretical programming languages and two mathematical formalisms which were introduced in two bachelor degree computer science courses. The restriction to these languages was made so as to yield datasets which contained most, if not all, interactions of the students with these languages, and to ensure that the data would be less biased by the students' previous experiences: Common programming languages could be used by all students without the editor evaluated in this article, with many of them having previous experiences with the languages taught or similar programming languages.

The evaluation reported about in this article is driven by the following research questions:

1. Does the editor's usage correlate with the examination results?
2. Are there patterns of behaviour in the use of the editor which are typical of students succeeding (or failing, respectively) at the examination?
3. Are there patterns of behaviour in the editors' use which are typical of students using the editor very shortly or rarely?
4. What is the students attitude towards the editor?

The tool was integrated into the teaching practice: Students could, and were encouraged to, work out homework assignments using the editor. In total, usage data of 102 students gathered over two semesters were evaluated.

Research question one was answered positively, counts of active usage phases significantly correlated positively with the examination results ($r = 0.25$).

Research question two was also answered positively: Students that used the tool both for examination preparation and for exercises during the semester were more successful in the final examination than students that did not use the tool at all or only in one of the two phases. Furthermore, students who received grades above average committed significantly fewer syntax errors while using the tool, and, interestingly, edited their code significantly more often before re-testing it than their less successful peers.

To answer research question three, the editor usage of students who used it more than twice and who are referred to as "fiddling students" was compared to the editor usage of the students who used the tool only once or twice. The following particular patterns of behaviour was observed among the "fiddling students": They often started their active phases with code containing examples or parts of examples provided with the homework assignments (which were syntactically correct but did not solve their problem), had longer active phases, and modified their code more before retesting it. Their "non-fiddling" peers on the other hand made significantly more syntax errors which occurred earlier within their active phases, and more often tried to re-test their code without modifying it.

The students' reception of the editor was generally positive and students reported that they considered the editor to be beneficial for their learning.

This article is structured as follows: Sect. 2 discusses related work, Sect. 3 describes the learning tool and the gathered data in detail, Sect. 4 presents the results, and Sect. 5 concludes this article by discussing the results and providing perspectives for future work and implications for STEM education.

2 Related Work

The following section discusses related work by briefly establishing connections to the Discovery Learning theory, and the model of programming skill acquisition by Lopez et al. as results of both these theories were in part reproduced by the research presented in this article. Finally, educational software similar to the editor evaluated in this article are discussed.

Discovery Learning for Formal Languages. The work presented in this article is related to Discovery Learning which is defined as "a type of learning where learners construct their own knowledge by experimenting with a domain, and inferring rules from the results of these experiments" [16], the "domain" being in this case formal languages. To the authors' best knowledge "Discovery Learning" to learn formal languages is not a thoroughly investigated field, yet computer science was reported to be "predestined for Discovery Learning" [2] and approaches of Discovery Learning has been successfully deployed in computer science education [13]. Important to Discovery Learning are cognitive tools [16], of which the editor presented in this article is an example: This editor aims at reducing the students' mental work (for instance by executing a `LOOP`-program instead of letting the students argue about the programm's inner working).

Programming Skill Acquisition. Difficulties in the acquisition of programming skills are often related to insufficient skills in problem solving [10,11,15]. Yet, Lister et al. identified problems (by using fill-in-the-blank and similar exercises) that were not directly related to problem solving, yet skills that had to be mastered before problem solving could take place [8]. Based on these experiments, Lopez et al. provided and validated a hierarchical model of programming skill acquisition, the lowest level being "Basics" which encompasses the recognition of syntax errors [9]. Much of the evaluation presented in this article can be related to this basic level. Other authors argue that many problems encounter while programming is caused by the necessity of mastering simultaneously syntax and semantics of the language [4,6].

Web-Based Programming and Pedagogical Languages. Much research has been conducted on supporting programming [1,7,14], constructing logical proofs [5] and manipulating algebraic terms [17] in the browser. Web-based systems are often employed as they provide an easy access for beginners and often ease the assessment of student submissions [7,12].

Simulating machines and letting students control the operation code (which is typically specified in a formal language) has been shown to be a very successful teaching method in computer science and physics [3,18].

3 Methods and Dataset

The following section describes the evaluation method, the integration of the editor in the teaching routine, and the evaluated dataset. The research method employed consists of gathering usage data during a course in which the editor is offered non-compulsory, and examining that data for inherent patterns and patterns related to examination success. This method allows to observe self-directed learning over a relatively long time period spanning from weeks to months which would hardly be possible in a laboratory setting.

Courses, Participants, and Organization. The editor was deployed in two bachelor degree computer science courses. The first course, an introduction to theoretical computer science, took place in the summer semester of 2018, lasting 14 weeks from April to July 2018. It was attended by 433 students of whom 113 were female and 315 male. 210 Students attended the final examination, 177 delivered at least one homework exercise, and 85 students used the editor at least once. The editor supported this course with five formal languages: Control codes for push-down-automata and Turing Machines, which could be specified and tested in the editor (See Fig. 1, right), and three imperative programming languages LOOP, WHILE and GOTO, each providing different means to realize repetition:

- LOOP: repetition can be expressed by so-called for loops the number of which is fixed in advance.
- WHILE: repetitions can be expressed by so-called while loops the number of which can be changed within the loops.
- GOTO: repetitions can be expressed by unconditional branching, using so-called go to branching instructions.

For each language, homework exercises were part of the course. The second course, a course on discrete mathematics, took place in the winter semester of 2018 lasting from September 2018 to February 2019. It was attended by 42 students, 17 of which used the tool. At the time of the preparation of this article, the final examination of this course was not yet conducted, and could therefore not be included in this evaluation. The tool supported this course with two mathematical formalisms: A scaffold for transforming expressions of predicate logic into clause normal form (See Fig. 1 right), and a scaffold verifying proofs by structural induction.

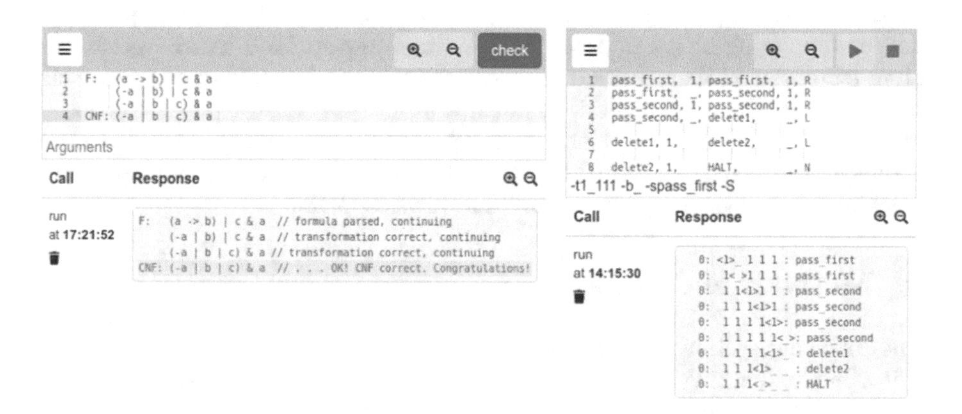

Fig. 1. Examples of supported languages. Left: successful transformation of a logical expression into clause normal form. Right: simulation of a turing machine.

Data Collection and Processing. Each time students tested (i.e. automatically verified or executed) their code, an activity was registered consisting of the tested source code and the time at which the test was performed. An active phase consists of a sequence of such activities, where two subsequent activities lie less than two hours apart. For each activity (except for the first in a phase), the Levenshtein distance of this activities' code to its preceding activities' code was computed. Note that the Levenshtein distance (also referred to as edit distance) is the minimum number of operations needed to transform one text (in this case source code) into another. Of special interest was the edit distance 0, expressing that no modifications were made between tests. Average edit distances, phase durations, and activity counts within a phase were computed for each student.

The proportion of activities producing syntactically correct code was computed for each student, as well as the average point of the first syntax error within an a students' active phases. Finally, all students could use and modify code examples and templates that were part of the courses' learning material provided by the teachers. To determine whether these examples and templates were used, the student code was automatically searched.

4 Results

In the following, results of several statistical hypothesis tests are presented. A significance threshold of $p = 5\%$ was chosen.

Usage Patterns Related to Examination Success. The count of active phases of a student correlated significantly with his or her examination mark (Pearson correlation test, $r = 0.26$). The activity count showed a weaker correlation ($r = 0.14$) with the examination mark, which indicates that the measure of active phases is sensible for the following evaluation.

Two-sided t-tests were conducted to compare students with above and below average marks. Students with above average marks made significantly fewer syntax errors, and edited their code more before testing it.

Two kinds of active phases were observed. The first kind took place during the semester, usually shortly after a regarding exercise was published, resulting in notable activity peaks. The second kind took place shortly before the examination which resulted in a final activity peak. Figure 2 shows a histogram of activities of the Turing machine simulator which could be used to work out two homework assignments with notable peaks of activity before the topics introduction, the homework deadlines and the final examination. Students showing both kind of activities were most successful, students showing only the first kind of activity were slightly less successful, and students who showed only the second kind of activity, or did not use the tool at all were significantly less successful than students in the two other groups (see Fig. 2 right).

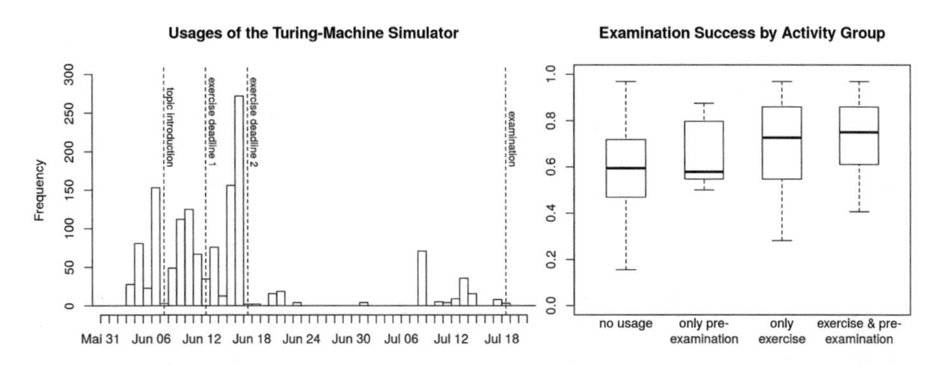

Fig. 2. Right: examination success of different groups of students. Left: activities of the turing-machine simulator from introduction of the topic, two homework deadlines to the final examination.

Usage Patterns Related to Activity Counts. The tool usage was very unequally distributed: 35% of the students performed 90% of the absolute activity, with 61% of the students using the tool only once or twice. These "non-fiddling" students are of special interest, as they could obviously not benefit from the tool. Two sided t-tests were conducted to test for differences between these "non-fiddling" and "fiddling" students. Non-fiddling students made significantly less use of the examples and templates provided than the fiddling students. They had a significantly higher frequency of syntax errors, which happened earlier in their active phases, which, in turn, were significantly shorter. Furthermore they edited less between tests, and re-tested the same code without modifying it more often. To further investigate the pattern of "re-testing code without changing it", the regarding code samples were examined; all of them contained syntax errors.

Student Attitudes Towards the Tool. 27 students completed a survey in August 2018 on the usability of the learning platform the editor is embedded in, and on the editor in particular. The students were asked to indicate on a six point Likert scale ranging from "not at all" (1) to "absolutely" (6) the perceived helpfulness, the ease of use, and the ease of use to test code provided by other users. The helpfulness was rated with an average of 4.4 (variance 1.9), the ease of use was rated with an average of 3.4 (variance 1.9) and the ease of testing foreign code with 2.4 (variance 2.5), as indicated by Fig. 3.

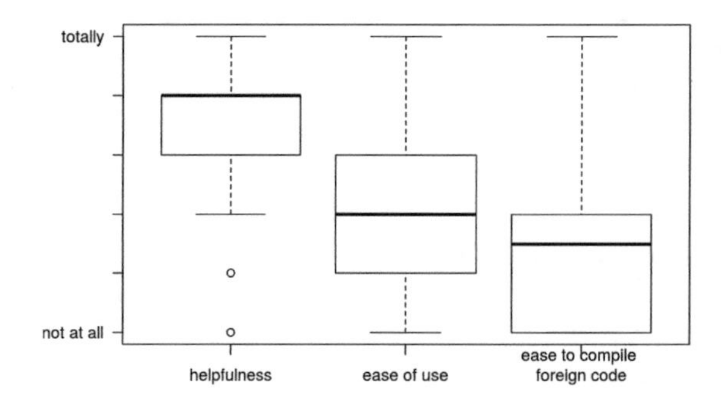

Fig. 3. Survey results regarding the helpfulness, the ease of use and the ease to test foreign code.

13 students mentioned the editor in positive statements like "The language compiler/interpreter is a really cool feature" or "The online compilation tools are somewhat useful." Better error messages were mentioned as a main point for improvement.

5 Discussion and Perspectives for Future Work

The survey results indicate that the usability could be improved, and efforts should be undertaken to let students try out their peers code more easily. However, the correlation of editor usage and examination results, the reported helpfulness of the editor, and the fact that many successful students reused the editor for their examination preparation indicates that the students benefited from it.

With regard to this finding, the majority of "non-fiddling" students, who could not benefit from the editor and used it only rarely is striking. A main difference in behaviour was identified to be the frequency of syntax errors they committed. Possibly, and despite of the languages having been introduced in class, these students could not "get the tool to work" and abandoned it after a few tries. The pattern of "re-testing unchanged code" is interesting: It is an irrational behaviour (code will not just start working on its own) and may be a further sign of frustration. Interestingly, neither questions on the editor or on the error messages it delivers were posed on the learning platform during the courses the editor is embedded in.

"Fiddling" students, constituting a minority performing the majority of activities, relied more often on code the teaching staff provided. Quite possibly, these students were not more adept in the beginning, but modified (or "fiddled with") working examples until it fit their needs, and, as they were less frustrated by the syntax, benefited more from the tool.

Based on these conclusions, the following recommendations can be made: Firstly, "fiddling" could be integrated into the teaching routine. Exercises introducing a new formal language could constitute of analysing or modifying given examples, instead of asking to produce working code. In this way, the desired "fiddling" behaviour could be encouraged. Secondly, attempts should be made to reduce frustration (and therefore attrition). Error messages should be informative an precise, possibly even referring to additional resources, instead of simply pointing to errors. Software could analyse learner behaviour, especially detecting sequences of activities that fail due to syntax errors, or re-tests of non-working code. In such a situation, software could suggest a correct example or establish contact between the student and a tutor or peer.

This article reported on an evaluation of an online learning tool to support the leaning of formal languages in two bachelor degree computer science courses. The results indicate that the tool helped some students in their learning. Interesting patterns of behaviours, referred to as "code fiddling", were identified and related to student success in the final examination. Suggestions for teaching STEM fields, which rely heavily on the use of formal languages were made.

Acknowledgement. The authors are thankful to Norbert Eisinger, Elisabeth Lempa, Marinus Enzinger, Caroline Marot and Thomas Weber for providing the interpreters evaluated in this article.

References

1. Amelung, M., Piotrowski, M., Rösner, D.: EduComponents: experiences in e-assessment in computer science education, vol. 38. ACM (2006)
2. Baldwin, D.: Discovery learning in computer science. In: ACM SIGCSE Bulletin, vol. 28, pp. 222–226. ACM (1996)
3. De Jong, T., Van Joolingen, W.R.: Scientific discovery learning with computer simulations of conceptual domains. Rev. Educ. Res. **68**(2), 179–201 (1998)

4. Gomes, A., Mendes, A.J.: Learning to program-difficulties and solutions. In: Proceedings of the International Conference on Engineering Education–ICEE, vol. 2007 (2007)
5. Hendriks, M., Kaliszyk, C., Van Raamsdonk, F., Wiedijk, F.: Teaching logic using a state-of-the-art proof assistant. Acta Didactica Napocensia **3**(2), 35–48 (2010)
6. Jenkins, T.: On the difficulty of learning to program. In: Proceedings of the 3rd Annual Conference of the LTSN Centre for Information and Computer Sciences, vol. 4, pp. 53–58. Citeseer (2002)
7. Kaya, M., Özel, S.A.: Integrating an online compiler and a plagiarism detection tool into the moodle distance education system for easy assessment of programming assignments. Comput. Appl. Eng. Educ. **23**(3), 363–373 (2015)
8. Lister, R., Adams, E.S., Fitzgerald, S., Fone, W., Hamer, J., Lindholm, M., McCartney, R., Moström, J.E., Sanders, K., Seppälä, O., et al.: A multi-national study of reading and tracing skills in novice programmers. In: ACM SIGCSE Bulletin, vol. 36, pp. 119–150. ACM (2004)
9. Lopez, M., Whalley, J., Robbins, P., Lister, R.: Relationships between reading, tracing and writing skills in introductory programming. In: Proceedings of the Fourth International Workshop on Computing Education Research, pp. 101–112. ACM (2008)
10. McCracken, M., Almstrum, V., Diaz, D., Guzdial, M., Hagan, D., Kolikant, Y.B.D., Laxer, C., Thomas, L., Utting, I., Wilusz, T.: A multi-national, multi-institutional study of assessment of programming skills of first-year CS students. In: Working Group Reports from ITiCSE on Innovation and Technology in Computer Science Education, pp. 125–180. ACM (2001)
11. Perkins, D., Martin, F.: Fragile knowledge and neglected strategies in novice programmers. In: First Workshop on Empirical Studies of Programmers on Empirical Studies of Programmers, pp. 213–229 (1986)
12. Pritchard, D., Vasiga, T.: Cs circles: an in-browser python course for beginners. In: Proceeding of the 44th ACM Technical Symposium on Computer Science Education, pp. 591–596. ACM (2013)
13. Ramadhan, H.A.: Programming by discovery. J. Comput. Assist. Learn. **16**(1), 83–93 (2000)
14. Rodríguez, S., Pedraza, J.L., Dopico, A.G., Rosales, F., Méndez, R.: Computer-based management environment for an assembly language programming laboratory. Comput. Appl. Eng. Educ. **15**(1), 41–54 (2007)
15. Rohwer, W.D., Thomas, J.W.: The role of autonomous problem-solving activities in learning to program. J. Educ. Psychol. **81**(4), 584 (1989)
16. Van Joolingen, W.: Cognitive tools for discovery learning. Int. J. Artif. Intell. Educ. (IJAIED) **10**, 385–397 (1998)
17. Virvou, M., Moundridou, M.: A web-based authoring tool for algebra-related intelligent tutoring systems. Educ. Technol. Soc. **3**(2), 61–70 (2000)
18. Vollmar, K., Sanderson, P.: MARS: an education-oriented MIPS assembly language simulator. In: ACM SIGCSE Bulletin, vol. 38, pp. 239–243. ACM (2006)

Can Text Mining Support Reading Comprehension?

Eliseo Reategui[1(✉)], Daniel Epstein[1], Ederson Bastiani[1], and Michel Carniato[2]

[1] PPGIE, Federal University of Rio Grande do Sul (UFRGS), Porto Alegre, Brazil
eliseoreategui@gmail.com, daepstein@gmail.com, edersonbastiani@gmail.com

[2] Pontifícia Universidade Católica do Rio Grande do Sul (PUCRS), Porto Alegre, Brazil
michelcarniato@hotmail.com

Abstract. Text mining is a research field that has developed different techniques to find relevant information in unstructured data, such as texts. This article tries to verify whether the automatic extraction of information from texts can help students in reading comprehension activities. Two studies involving control and experimental groups were carried out with students of 5th and 8th grade in order to evaluate whether a particular text mining tool could effectively help students improve their scores in a reading task. We also wanted to verify if the use of the tool by students in different grades could yield different outcomes. Results showed that the mining tool helped 5th graders to improve their scores, but it was not so effective for 8th graders. These results indicate the potential of the proposed tool especially for learners who are still developing their reading skills.

Keywords: Reading comprehension · Text mining · Graphic organizers · Literacy

1 Introduction

For years, researchers have been debating over fact that young students tend to read less, drawn by other stimuli such as television and video games [31]. The Internet also changed drastically the way we interact with information and the written word, especially as we often have access to information through audio and video without the actual need to read [29]. Furthermore, when accessing information in the Internet, more and more students get used to skimming, reading titles and fragments without taking their time to think and reflect about what was written. Tovani [28] calls this ability "fake reading", which gives people a faint idea of what is written, but does not really allow them to remember or retell what they read in more detail. However, today's expectations in terms of literacy involve the ability to participate in the meanings of text, to use texts functionally and to critically analyze and transform texts [30]. In this contemporary vision, individuals are expected to use oral and written language

R. Gennari et al. (Eds.): MIS4TEL 2019, AISC 1007, pp. 37–44, 2020.
https://doi.org/10.1007/978-3-030-23990-9_5

demonstrating an understanding of the world, being able to communicate, to participate in problem solving and decision making [14]. In a digital communication age, reading and writing has also assumed a new role in the way people interact, share information and socialize [26].

Regarding the process of reading comprehension, it can be thought of as a hierarchical process of actions to identify letters, words and sentences to give meaning to what is read [27]. In this context, reading itself is only part of the process, which involves actions such as decoding symbols, identifying phonemes and semantic associations. Reading comprehension requires the reader to make associations and give meaning to the collection of letters and symbols that compose a particular text [4, 6]. These actions occur before, during, and after reading. Beers [2] stresses that it is necessary to explicitly and systematically approach all aspects of textual comprehension, because focusing on only one of them is not enough to ensure that the reader develops this ability successfully. In this sense, the American National Reading Panel [18] identified five core components that are essential in reading program: phonological awareness, phonics, vocabulary, comprehension, and fluency. Reading interventions usually focus on one or more of these components, as remarked in a literature review done by Jamshidifarsani et al. [13]. Furthermore, the types of technology proposed to support reading activities are quite varied, from adaptive learning systems that support story reading and question/answering activities [7], to summary extraction methods to help learners with reading difficulties [17]. In this project, our main research question has been to evaluate how text mining could be used to help students in reading comprehension tasks. The idea has been to use the power of text mining to identify important terms and relationships in texts, and provide this information to students as a starting point for the text analysis. Sobek text mining has been the tool chosen for the proposed tasks. Sobek is capable of extracting relevant terms from a text and representing them in a graph, using a particular mining algorithm based on the n-simple distance graph model, in which nodes represent the main terms found in the text, and the edges describe adjacency information [24]. The next section presents related work.

2 Related Work

Previous research has shown how nonlinguistic representations can help students in reading and writing tasks. Marzano, Pickering and Pollock [16], for example, discussed the importance of using nonlinguistic representations to help students enhance their understanding of written material. Hyerle [12] showed how different types of visual tools, called graphic organizers, could help students and teachers represent information and communicate with others. Concept maps are a type of graphic organizers that have been used extensively to help students in learning tasks. For instance, Guastello, Beasley and Sinatra [8] showed that concept maps could help low-achieving 7th graders to improve their scores in science exams. Romero et al. [23] also showed how concept maps could help students between 13 and 14 in learning tasks in a Natural Sciences course, providing advantages such as the agility to organize concepts in a new domain and the possibility to visualize ideas and relationships in a simple way. It is possible to find a large body of research about the use of concept maps to help students in learning activities in different knowledge areas, as well as in different educational levels [19]. However, no matter how simple the construction of a concept map may be, it hasbeen shown that even college students have difficulties to build and revise a concept map [20]. Therefore, providing the students with an initial draft they can work with can be a helpful approach, which is the proposal presented in this article. We use text mining to give the student a rough representation of terms and relationships between them that may be important to consider. Then, the students have to manipulate these graphic representations to make them closer to their own understanding of the read. The next section presents Sobek, a tool which has been developed to build graphs from texts, using the technique of text mining.

3 Text Mining and Sobek Mining Tool

Text mining is a research field that encompasses different approaches such as information retrieval, natural language processing, information extraction, text summarization, supervised and unsupervised learning, probabilistic methods, text streams and social media mining, opinion mining and sentiment analysis [1]. In the field of Education, text mining has become more popular especially with the advent of distance learning and Massive Open Online Courses [25]. It has been used, for example, in the analysis of student online interaction, showing that the combination of text and data mining, applied to a large data set, can reveal relevant information about students' behavior [10]. Text mining has also been used to conduct formative assessment and let learners and instructors to visualize results, providing an alternative solution to evaluate learners' performance throughout the learning process [11]. Researchers have also evaluated how the mining of student responses from a survey could yield relevant information for management purposes [33], and contrasted the results obtained through the mining of students' opinions about teacher leadership with those of human raters [32].

In this research, we have focused specifically on the use of a particular text mining tool (Sobek) to support reading comprehension, in an attempt to help students overcome their difficulty to delve into a deeper level of reflection when they read. Sobek provides a graphical representation of any text, based on the concepts extracted from it. Sobek's operation can be divided into three stages. The first one consists in identifying the relevant concepts in the text and summarizing them. The second step is related to the identification of relationships among those concepts, and the last one concerns the visual representation of the information extracted in the form of a graph. These graphs can be thought of as graphic organizers, which have been used extensively to help students and teachers represent information and communicate with others [12]. Our focus in this research has been to use Sobek to extract from the text a visual representation of relevant terms, and ask students to analyze and modify it according to their own understanding of the text. Further detail about Sobek's text mining algorithm and operation may be found in [21].

4 Methods

Two studies involving control and experimental groups were carried out with a total of 98 students of 5th and 8th grade in order to evaluate whether Sobek could effectively support reading comprehension activities. We also wanted to verify whether the use of the mining tool by students with different reading abilities could yield different results. Participants were members of a project called "Social Action", which focused on providing better learning opportunities to low-income students. All of the participants were familiar with using a computer. The reading comprehension tests used in the experimental studies were composed of texts and questions taken from a standard national literacy assessment. There was no time limit for the activity, students could take as long as they needed to read the texts and answer the questions.

The first experiment was carried out with 51 students of 5th grade, 24 of them in the control group, in which Sobek was not used, and 27 in the experimental group. While the control group did the activity using the computer to read the texts and answer the questions, the experimental group was able to use Sobek during the activity, following the instructions below:

- Read the text
- Copy the text into Sobek
- Use the mining tool to extract a graph from the text
- Add terms that you consider to be relevant but that are not in the graph
- Remove from the graph terms that you consider to be irrelevant
- Add/remove relationships between terms that you understand to be important
- Answer the questions about the text

Students of the experimental group were not instructed during the activity to make changes in their graphs as to make them smaller or larger in any way. On the contrary, they were left free to use the tool firstly to identify automatically an initial set of terms, and then to edit these terms/connections to create their own representation/ interpretation of the text. Figure 1 shows an example of graphs built by 5th grade students for a text called "The soccer-ball owner", by Ruth Rocha (originally in Portuguese). On the top of the figure we see the graph with the terms extracted by Sobek automatically from the text. On the bottom left we see the graph modified by student A, who included the words soccer and drill. On the bottom right we see the work of student B who also incorporated in his graph the word soccer, but included other terms too: *Caloca* (Carlos Alberto's nickname), *Captain, friends*. Neither of the two students removed any term from Sobek's original graph.

The identification of relevant terms/ideas in the text and then structuring them in a graphic organizer is one of several reading/comprehension strategies that can be taught to students in different educational stages [22]. When students work on a visual representation of the main ideias of a text, they have the opportunity to activate their previous knowledge and make connections to what they read, which leads to a better understanding of the text. In this sense, there is no right or wrong way to edit Sobek's graphs. Our approach to reading comprehension is more related to the effort the student

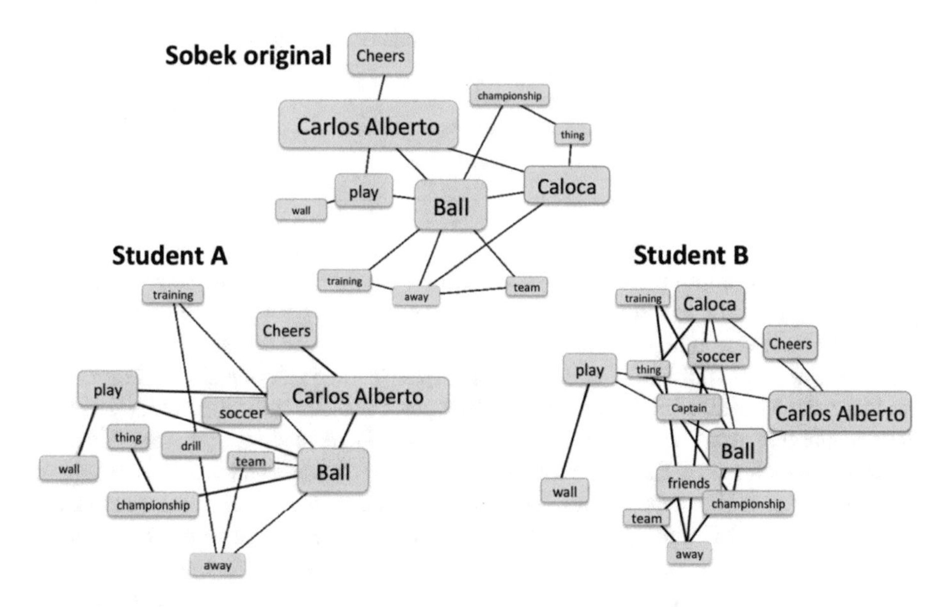

Fig. 1. Sobek and students' graphs for the text *"The soccer-ball owner"*

has to put into making sense of the visual representation extracted from the text and building their own associations and interpretations about what they read.

The second study had exactly the same format as the one described above, but it was carried out with students from two 8th grade classes, totalling 43 students, 23 male and 20 female. The students were divided in experimental and control group, 27 of them in the experimental group and 14 in the control group. The students were given a standardised test suited to their grade and therefore different from the one used with the 5 graders. Results from both studies are presented in Table 1.

Table 1. Results for experimental and control group of the 5th and 8th grade classes

Grade	Condition	Number of students	Performance (0.0–10.0)
5th	Control	24	4.80
5th	Experimental	27	6.20
8th	Control	16	4.70
8th	Experimental	12	6.51

In order to verify the statistical difference between the control and experimental group, Pearson's correlation coefficient was used. We computed the correlation between the use of the Sobek and the performance of the students. Table 2 presents the results of the statistic analysis.

For the 5th grade students the correlation coefficient was 0.48. This value indicates that there was a positive correlation between Sobek's use and the students' performance, with a *p-value* lower than 0.05, which indicates statistical significance.

Table 2. Pearson's correlation coefficients

Grade	Correlation between correct answers and Sobek's use	*p-value*
5th	0.48	0.00035
8th	0.36	0.087
p-value limit = 0.05		

Therefore, we could conclude that Sobek helped the 5th graders in the reading comprehension task. For the 8th grade students, the average performance of the experimental group was also higher than that of the control group. In this study, however, this difference was not statistically significant (*p-value* $\geq$ 0.05). A plausible interpretation for these results is that Sobek provided less support for older students. In this sense, we have to observe that 8th graders have more consolidated reading and comprehension strategies, which means that the introduction of novel methods and artefacts to support their current reading methods are less effective. These results reinforce Manoli and Papadopoulou's [15] findings that apprentice readers may benefit more from graphic organizers in learning activities than students who are accomplished readers.

Sobek was built with a log system that kept track of all the actions done by the user when editing the graphs. This analysis showed that 8th grade students made more changes in the graphs than those of the 5th grade. The older students also started the editing process by removing nodes, which was different from the students of 5th grade who were more reluctant to remove terms from the graph. The students of 8th grade also made more changes in the graphs' connections, being more confident in the identification of relationships between different pieces of information. This more determined attitude of 8th grade students also indicated that they were more assertive about what they read and understood of the text, so that they did not benefit so much from the analysis and manipulation of the graphs. Still, further experimental studies comparing younger and older students could provide clearer evidence about the benefits and limitations of the text mining and graph editing approach to support reading and comprehension activities.

5 Conclusion

This article presented two experimental studies about the use of text mining with visual representation of relevant terms to help students in reading comprehension activities. The studies, carried out with 5th and 8th grade students, demonstrated that the mining tool improved 5th graders scores in a reading activity, but was not so effective for 8th graders. These results indicate the potential of the proposed tool for learners who have not developed their complete mastery of reading skills, which is aligned with the idea that concept mapping is a strategy that is better suited to elementary students [15]. The use of text mining may provide relevant cues about topics in reading material that are likely to be important and may help students in reading comprehension tasks. In times when students are inclined to read quickly smaller fragments of text, it becomes important to create effective reading strategies that make them delve into deeper thinking processes. Previous research results have shown that graphs are a suitable way to represent information extracted from texts because of their simple organization and interpretation, in which nodes represent concepts/ideas and connections represent relationships between them [5]. These results support the decision to use Sobek's graphs as a starting point for the students to build their graphic organizers representing the main ideas of texts.

For future work, we are trying to understand how students may benefit from using Sobek in reading activities that involve group discussion and collaborative construction of visual representations of texts. Another important path is to understand if the reading strategies developed with the use of Sobek are incorporated by readers, little by little, and if later they can perform well in reading comprehension tasks without the use of the mining tool.

References

1. Allahyari, M., et al.: A brief survey of text mining: classification, clustering and extraction techniques. In: Proceedings of Conference on Knowledge Discovery and Data Mining - KDD, Halifax, Canada (2017)
2. Beers, K.: When Kids Can't Read. What Teachers Can Do. Heinemann, Portsmouth (2003)
3. Block, C.C., Pressley, M.: Comprehension Instruction: Research-Based Best Practices. Solving Problems in the Teaching of Literacy. Guilford Press, New York (2002)
4. Chang, K.E., Sung, Y.T., Chen, S.F.: Learning through computer-based concept mapping with scaffolding aid. J. Comput. Assist. Learn. **17**(1), 21–33 (2001)
5. Chein, M., Mugnier, M.-L.: Graph-Based Knowledge Representation: Computational Foundations of Conceptual Graphs, 1st edn. Springer, London (2008)
6. Cordón, L.A., Day, J.D.: Strategy use on standardized reading comprehension tests. J. Educ. Psychol. **88**(2), 288–295 (1996)
7. Di Giacomo, D., Cofini, V., Di Mascio, T., Rosita, C.M., Fiorenzi, D., Gennari, R., Vittorini, P.: The silent reading supported by adaptive learning technology: influence in the children outcomes. Comput. Hum. Behav. **55**(1), 1125–1130 (2016)
8. Guastello, E.F., Beasley, T.M., Sinatra, R.C.: Concept mapping effects on science content comprehension of low-achieving inner-city seventh graders. Remedial Spec. Educ. **21**(6), 356–364 (2000)
9. Hall, T., Strangman, N.: Graphic organizers. National Center on Accessing the General Curriculum, Wakefield (2002)
10. He, W.: Examining students' online interaction in a live video streaming environment using data mining and text mining. Comput. Hum. Behav. **29**(1), 90–102 (2013). https://doi.org/10.1016/j.chb.2012.07.020
11. Hsu, J.-L., Chou, H.-W., Chang, H.-H.: EduMiner: using text mining for automatic formative assessment. Expert Syst. Appl. **38**(4), 3431–3439 (2011)
12. Hyerle, D.: Visual Tools for Transforming Information Into Knowledge. SAGE, Thousand Oaks (2008)
13. Jamshidifarsani, H., Garbaya, S., Lim, T., Blazevic, P., Ritchie, J.M.: Technology-based reading intervention programs for elementary grades: an analytical review. Comput. Educ. **128**(1), 427–451 (2019)
14. Jenner, J.: A bridge to reading and writing literacy: developing oral language skills in young children. Pacific Educator (2003)

15. Manoli, P., Papadopoulou, M.: Graphic organizers as a reading strategy: research findings and issues. Creative Educ. **3**, 348–356 (2012). https://doi.org/10.4236/ce.2012.33055
16. Marzano, R.J., Pickering, D.J., Pollock, J.E.: Classroom Instruction that Works: Research-Based Strategies for Increasing Student Achievement. Association for Supervision and Curriculum Development, Alexandria (2001)
17. Nandhini, K., Balasundaram, S.R.: Improving readability through extractive summaries for learners with reading difficulties. Egyptian Inform. J. **14**(1), 195–204 (2013)
18. National Reading Panel: Teaching children to read: an evidence-based assessment of the scientific research literature on reading and its implications for reading instruction. NIH Publication. No. 00-4769 (2000) https://doi.org/10.1002/ppul.1950070418
19. Nesbit, J.C., Adesope, O.O.: Learning with concept and knowledge maps: a meta-analysis. J. Rev. Educ. Res. **76**(3), 413–448 (2006)
20. Reader, W., Hammond, N.: Computer-based tools to support learning from hypertext: concept mapping tools and beyond. Comput. Educ. **12**, 99–106 (1994)
21. Reategui, E., Epstein, D., Lorenzatti, A., Klemann, M., Sobek: a text mining tool for educational applications. In: International Conference on Data Mining, Las Vegas, USA, pp. 59–64 (2011)
22. Roman-Sanchez, J.M.: Self-regulated learning procedure for university students: the meaningful text-reading strategy. Electron. J. Res. Educ. Psychol. **2**(1), 113–132 (2004)
23. Romero, C., Cazorla, M., Buzón, O.: Meaninful learning using concept maps as a learning strategy. J. Technol. Sci. **7**(3), 313–332 (2017)
24. Schenker, A.: Graph-theoretic techniques for web content mining. Ph.D. thesis, University of South Florida, Tampa, FL (2003)
25. Shatnawi, S., Gaber, M.M., Cocea, M.: Text stream mining for massive open online courses: review and perspectives. Syst. Sci. Control Eng. **2**(1), 664–676 (2014)
26. Sweeny, S.M.: Writing for the instant messaging and text messaging generation: using new literacies to support writing instruction. J. Adolesc. Adult Literacy **54**(2), 121–130 (2010)
27. Tankersley, K.: The Threads of Reading: Strategies for Literacy Development. ASDC, Alexandria (2003)
28. Tovani, C.: I Read But I Don't Get It. Comprehension Strategies for Young Adolescent Readers. Stenhouse Publishers, Portsmouth (2000)
29. Trend, D.: The End of Reading: From Gutenberg to Grand Theft Auto. Peter Lang Publishing, New York (2010)
30. Warschauer, Mark: Laptops and Literacy: Learning in the Wireless Classroom. Teachers College Press, New York (2006)
31. Wiecha, J.L., Sobol, A.M., Peterson, K.E., Gortmaker, S.L.: Household television access: associations with screen time, reading and homework among youth. Ambul. Pediatr. **1**(5), 244–251 (2001)
32. Xu, Y., Reynolds, N.: Using text mining techniques to analyze students' written responses to a teacher leadership dilemma. Int. J. Comput. Theory Eng. **4**(4), 575 (2012)
33. Yu, C.H., DiGangi, S.A., Jannasch-Pennell, A.: Using text mining for improving student experience management in higher education. In: Tripathi, P., Mukerji, S. (eds.) Cases on Innovations in Educational Marketing: Transnational and Technological Strategies, pp. 196–213. Information Science Reference, Hershey (2011). https://doi.org/10.4018/978-1-60960-599-5.ch012

Usability of Virtual Environment for Emotional Well-Being

Elisa Menardo(✉), Diego Scarpanti, Margherita Pasini, and Margherita Brondino

University of Verona, Verona, VR, Italy
{elisa.menardo, diego.scarpanti, margherita.pasini, margherita.brondino}@univr.it

Abstract. Nature is nowadays considered the most restorative environment. A brief exposure to a (real or virtual) nature environment help people to restore cognitive and affective resources. However, (1) the precise mechanism through nature support restorative process is undervalued in literature; (2) testing the possibility that a virtual reality (VR) device can evoke similar outcomes as real nature, the usability of such devices become an important factor to consider in the research. In this study we hypothesize a moderation effect of usability of device on the relationship between perceived restorative potential (PRP) of the environment and restorative outcomes (emotional and cognitive). 114 Italian students (83% female, mean (ds) age = 22.38 (6.50)) were immersed into a virtual nature environment using a virtual reality head-set (*Oculus Rift*). Before and after exposure, mood (positive and negative emotions) and attentive performance were evaluated. After exposure participants were also asked to report PRP and usability of VR device experienced. Result showed that the PRP predict emotional well-being, and that this relationship is moderated by usability of VR device. Only participants that perceived a high usability report different levels of emotions depending on how much they perceived the environment restorative. Conversely, people how reported difficulties in wearing headset VR or to adapt to the environment (i.e., low level of usability) did not show difference in emotional well-being depending on PRP.

Keywords: Virtual reality · Emotional well-being · Restorativeness

1 Introduction

1.1 Background

An increasing body of researches suggest that nature is an important factor that should be implemented in the build environment [1]. Evidences from different fields of research, including psychiatric sciences, epidemiology, clinical and social health psychology, show how nature play a role in stress, mental fatigue and self-regulation failure [2]. A brief exposure to a natural environment may have a restorative effect on people's resources enhancing performance in cognitive tasks [1] and increasing positive mood [3]. Restorative environments are environments that promote, and not only allow, the recovery of resources (biological, cognitive, psychological, social) in an

R. Gennari et al. (Eds.): MIS4TEL 2019, AISC 1007, pp. 45–52, 2020.
https://doi.org/10.1007/978-3-030-23990-9_6

individual [4]. Two main theories have been proposed to explain why human beings benefit from natural environments: The Stress Recovery Theory (SRT) [5, 6] and the Attention Restoration Theory (ART) [7, 8]. Ulrich focuses on the pre-cognitive emotional response while Kaplan mainly on executive functioning aspect [2]. The restorative capability of an environment is often investigated by the estimation of its restorative potential (PRP) [9], based on the assumption that meta-cognitive abilities of individuals allow them to understand their cognitive processes and to estimate how they are influenced by different environments [10]. Literature on this field of research have highlighted that people perceive a higher restorative potential in nature than in urban environments [5, 6, 11–13].

However, "the extent to which perceived restoration predicts actual restoration is undervalued in the literature" [10, p. 2]. Marselle [14] suggest that the perceived restorativeness may mediate the impact of nature walk on resources depletion. The effect of nature on mood may depend on perceived restorative potential. This seems to be true also if people are exposed to a virtual nature environment [15], instead of a real nature environment [13], confirming the implicit assumption that exposure to simulated environments produces the same effects of exposure to real environments [16, 17]. However, even if experience in real and virtual environment have similarities [18], there are important difference, perhaps depending on the quality of the immersion in the simulated environment [19].

However, to our knowledge, only three studies have investigated the relationship between the perceived restorative potential (PRP) and the restorative impact of an environment on resources of an individual (restorative outcomes) [13–15], and all have considered only emotional outcomes. Secondly, the use of virtual reality device (e.g., Oculus Rift) in environmental psychology research is recent [e.g. 13, 16, 17] and no study have investigated his "usability" [20] in environmental psychology research. With the concept of usability, we consider not only what is easy to use, but also the concept of efficacy and effectiveness of human performance: "the capability in human functional terms to be used easily and effectively by the specified range of users, given specified training and user support, to fulfil the specified range of tasks, within the specified range of environmental scenarios" [20, p. 340]. According with Shackel [20], considering the four main components of human-machine interaction (user, task, tool, environment), the virtual reality may be conceptualized as a merging of the tool and the environment, where the tool become the environment and the environment is provided by the tool. Thus, the usability of this tool (VR environment) become a key factor in research testing the possibility that a virtual reality device can evoke similar outcomes as real nature. Credibility of the virtual environment is necessary to create the condition of sense of being in task environment, 'presence', spatial immersion, mirroring the real world [21].

1.2 The Present Study

The aim of this study is to investigate the usability of Oculus Rift and to verify its influence as a potential moderator on the relationship between virtual environment and the restorative outcomes. In particular, we hypothesize a moderation effect of usability of device on the relationship between PRP and restorative outcomes (emotional and cognitive).

2 Method

2.1 Participants and Procedure

114 undergraduate students (83% female, mean (ds) age = 22.38 (6.50), range = 18 – 51) of University of Verona participated to the study.

After obtaining informed consensus, participants were individually called to the laboratory. First, they completed the Achievement Emotions Adjective List (AEAL) [22] and Sustained Attention to Response Task (SART) [23]. Second, participants saw a 360-degree video filmed in a semi-manicure wood (see Fig. 1) wearing a virtual reality (VR) head-set (*Oculus Rift*) [24]. Third, they completed the Perceived Restorativeness Scale (PRS) [25] and the AEAL and SART again. Finally, participants completed a series items about usability of the device.

Fig. 1. An image of the environment in which participants were immersed using virtual reality.

Oculus Rift is composed of two lenses that project images onto two OLED screens (1080 × 1200 resolution) completely covering the view of users. Users are totally immersed in the virtual environment with a peripheral view similar of real life (110° of visual angle). It is also equipped with headphones that produce a 3D sound effect and a gyroscope that follows the movements of the head on the 4 axes of view (from top to bottom, from right to left), so users can look around during the experience [24].

2.2 Instrument

Achievement Emotions Adjective List (AEAL). The scale is composed by 30 items, three for each emotion, related to three positive activating emotions (enjoyment, hope, pride), two positive deactivating emotions (relief, relaxation), three negative activating emotions (anxiety, shame, anger), and two negative deactivating emotions (boredom, hopelessness). We excluded the items related to pride and shame, because they were not salient for the study. Before and after exposure, participants were asked to indicate how they felt evaluating how much each word described their feelings on a 7-point Likert scale (1 = *not at all* and 7 = *very much*). Negative items are reversed, so high score to the scale means lower negative emotions.

Sustained Attention to Response Task (SART). The test is composed by 240 stimuli (i.e., number from *1* to *9*), presented one at a time, and the task is to response, as quickly as possible, to all stimuli except the number "*3*". The test was run on a screen (size 1280 × 720) using the Psychology Experiment Building Language (PEBL) [26].

Perceived Restorativeness Scale (PRS). The scale comprises 11 items, based on a 11-point Likert scale from *1* (Not at all) to *11* (Very Much), assessing the individual perception of restorative qualities (i.e., fascination, being-away, coherence, and scope) of an environment [25].

Usability of the device. Three ad hoc items were built to assess how fast (item #1) and easy (item #2) it was to adapt to virtual environmental and how fast it was to adapt to VR head-set (item #3). All items were based on a 5-point Likert scale from *1* (very difficult/slowly) to *5* (very easy/quickly).

2.3 Data Analysis

Data analysis was carried out using R software [27] following a quantitative approach. Correlation analyses were done as preliminary investigation of relationship between variables. A series of hierarchical multiple regression model were conducted to test the hypothesis that device's usability moderates the relationship between PRP and positive emotion, negative emotion, or SART performance (correct response or response time). In the first step we included single predictors (PRP and usability) (Model 1); in the second step we added the interaction term (Model 2) created after centering predictors to avoid multicollinearity [28].

3 Results

As showed in Table 1, SART outcomes did not significantly correlated with PRS, so they were excluded in sub-sequent analyses.

Model 1 accounted for a significant amount of variance in positive ($R^2 = .213$, $F(2,109) = 14.77$, $p < .001$) and negative emotion ($R^2 = .135$, $F(2,111) = 8.66$, $p < .001$), however, only PRS had a significant effect (positive emotions: $\beta = .41$, $t = 4.61$, $p < .001$; negative emotion: $\beta = .28$, $t = 3.06$, $p < .01$).

The interaction effect significantly increased the amount of variance in positive ($\Delta R^2 = .045$, $\Delta F(1,108) = 6.48$, $p < .05$, $\beta = .21$, $t(108) = 2.55$, $p < .05$) and negative emotion ($\Delta R^2 = .031$, $\Delta F(1,110) = 4.10$, $p < .05$, $\beta = .18$, $t(110) = 2.03$, $p < .05$). Examination of the interaction plot showed that at medium and high level of PRP, usability level enhances the PRP's effect on positive and negative emotion. By contrast, high level of usability has a "detrimental effect" on mood in people who perceived low restorativeness. Participant that perceived high level of restorativeness reported more

Table 1. Descriptives and Pearson correlation coefficient between variables (n = 114).

Variables	PRS	Positive emotion	Negative emotion	SCR	SRT	Usability
PRS	–					
Positive emotions	.443***	–				
Negative emotion	.325***	.441***	–			
SCR	.083	.211*	.181	–		
SRT	−.020	.037	−.025	.593***	–	
Usability	.221*	.236*	.166	.201*	.041	–
M	7.82	4.88	6.45	.78	400.78	12.13
SD	1.59	.96	.46	.15	67.37	2.06

Note. PRS = Perceived restorativeness Scale; SCR = SART Correct Response rate (%); SRT = SART Response Time (seconds); *p < 0.05. **p < 0.01. ***p < 0.001.

emotional well-being (more positive emotion, less negative emotion) when usability is high (versus low usability). By contrast, participants that perceived low level of restorativeness reported more emotional when usability is low (versus high restorativeness) (Fig. 2).

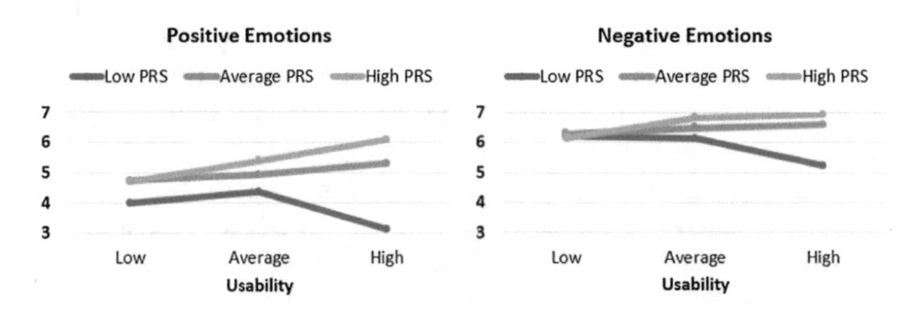

Fig. 2. Interaction plot for low (one *SD* below), average, and high (one *SD* above) level of Perceived restorative potential and usability on positive and negative emotions.

4 Conclusion

The aim of this study is to verify the potential influence of usability of Oculus rift on restorative outcomes after immersion into a virtual natural environment. Results suggest that a brief exposure to a virtual natural environment is enough to improve emotional well-being but not attentive performance. Moreover, perceived restorative potential of the environment is correlated with emotional well-being but not with attentive performance. These results could support hypothesis that restoration of emotional and cognitive resources depend on different mechanisms [29, 30]. However, the lack of the improvement in attention performance after exposure could be due to a too short time of exposure [17].

Moreover, results suggest that the perceived restorative potential predict emotional well-being, and that this relationship is moderated by usability of VR device. Indeed, when the perceived restorative potential of the environment is high, people that experience high usability report more emotional well-being that those who experience lower usability level. Conversely, when the perceived restorative potential of the environment is low people that experience high usability report lower emotional well-being that those who experience lower usability level. Only participants that perceived a high usability report different levels of emotions depending on how much they perceived the environment restorative. Conversely, people who reported difficulties in wearing headset VR or to adapt to the environment (i.e., low level of usability) did not show difference in emotional well-being depending on PRP.

In sum, these findings suggest that usability per se does not have an impact on emotional well-being, but it could influence relationship between well-being and its antecedents. So, we suggest that usability of the device is a key factor to consider, although when research want to study impact of a virtual environment on people's well-being, as well as degree of immersion [19]. VR is a flexible and economic technology that allow researchers to overcome time problems, physical inaccessibility and ethic problems [21]. For example, benefits consist in the chance to connect people at a distance, providing an interactive environment for learning [31] or in the possibility to use in a safe condition without real health risk for the patients [32].

For this reason, VR technology has been used in several fields as learning [31], training [33], and clinical [34]. Most of the studies had specific educational or training objective [21]. Advantages for all these contexts, especially in term of education, is the possibility to: (1) support the motivation with an involving environment; (2) support the access to event or person, as a teacher or expert, in distance; (3) avoiding dangerous situation (i.e. eliminating the possibility to make mistake in real environment); (4) taking into account ethic concerns: performing surgery without risk for real patients [21]. In all these applications usability may play an important role, with the level of visual definition or immersive capacity [18], to reach a threshold level of realism in the extent to provide a credible experience of the environment.

Future research should overcome limits of the present study considering, for example, different kind of environment with different level of naturalness, different kind of visual immersion, different level of interaction and different time of exposure, and different kind of cognitive tasks as a dependent variable.

References

1. Ohly, H., White, M.P., Wheeler, B.W., Bethel, A., Ukoumunne, O.C., Nikolaou, V., Garside, R.: Attention restoration theory: a systematic review of the attention restoration potential of exposure to natural environments. J. Toxicol. Environ. Health Part B **19**(7), 305–343 (2016)
2. Beute, F., de Kort, Y.A.: Salutogenic effects of the environment: review of health protective effects of nature and daylight. Appl. Psychol. Health Well-Being **6**(1), 67–95 (2014)
3. McMahan, E.A., Estes, D.: The effect of contact with natural environments on positive and negative affect: a meta-analysis. J. Posit. Psychol. (2015). https://doi.org/10.1080/17439760.2014.994224.

4. Hartig, T.: Restorative environments. In: Spielberger, C. (ed.) Encyclopedia of Applied Psychology, vol. 3. Academic Press, San Diego (2004)
5. Ulrich, R.S.: Visual landscapes and psychological weel-being. Landsc. Res. **4**(1), 17–23 (1979)
6. Ulrich, R.S.: Aesthetic and affective response to natural environment. In: Human Behavior and Environment, vol. 6: Behavior and Natural Environment. Plenum, New York (1983)
7. Kaplan, R., Kaplan, S.: The Experience of Nature: A Psychological Perspective. Cambridge University Press, New York (1989)
8. Kaplan, S.: The restorative benefits of nature: toward an integrative framework. J. Environ. Psychol. **15**, 169–182 (1995)
9. Staats, H.: Restorative environments. In: The Oxford Handbook of Environmental and Conservation Psychology. Oxford University Press, Oxford (2012)
10. Pearson, D.G., Craig, T.: The great outdoors? Exploring the mental health benefits of natural environments. Front. Psychol. **5**, 1–4 (2014)
11. Berto, R.: Assessing the restorative value of the environment: a study on the elderly in comparison with young adults and adolescents. Int. J. Psychol. **42**(5), 331–341 (2007)
12. Takayama, N., et al.: Emotional, restorative and vitalizing effects of forest and urban environments at four sites in Japan. Int. J. Environ. Res. Public Health **11**, 7207–7230 (2014)
13. Schutte, N.S., Bhullar, N., Richardson, E.J., Stilinovic, K.: The impact of virtual environments on restorativeness and affect. Ecopsychology **9**(1), 1–7 (2017)
14. Marselle, M.R., Irvine, K.N., Lorenzo-Arribas, A., Warber, S.L.: Does perceived restorativeness mediate the effects of perceived biodiversity and perceived naturalness on emotional well-being following group walks in nature? J. Environ. Psychol. **46**, 217–232 (2016)
15. McAllister, E., Bhullar, N., Schutte, N.S.: Into the woods or a stroll in the park: how virtual contact with nature impacts positive and negative affect. Int. J. Environ. Res. Public Health **14**(7), 786 (2017)
16. Valtchanov, D., Barton, K.R., Ellard, C.: Restorative effects of virtual nature settings. Cyberpsychol. Behav. Soc. Netw. **13**(5), 503–512 (2010)
17. Valtchanov, D., Ellard, C.: Physiological and affective responses to immersion in virtual reality: effect of nature and urban settings. J. CyberTher. Rehabil. **3**(4), 359–373 (2010)
18. de Kort, Y.A.W., IJsselsteijn, W.A., Kooijman, J., Schuurmans, Y.: Virtual laboratories: comparability of real and virtual environments for environmental psychology. Presence, Teleoperators Virtual Environ. **12**, 360–373 (2003)
19. de Kort, Y.A.W., Meijnders, A.L., Sponselee, A.A.G., Ijsselsteijn, W.A.: What's wrong with virtual trees? Restoring from stress in a mediated environment. J. Environ. Psychol. **26**, 309–320 (2006)
20. Shackel, B.: Usability-context, framework, definition, design and evaluation. Interact. Comput. **21**(5–6), 339–346 (2009)
21. Freina, L., Ott, M.: A literature review on immersive virtual reality in education: state of the art and perspectives. In: The International Scientific Conference eLearning and Software for Education Proceedings, p. 133, Bucharest (2015)
22. Raccanello, D., Brondino, M., Pasini, M.: Achievement emotions in technology enhanced learning: development and validation of self-report instruments in the italian context. Interact. Des. Arch. J. **23**, 68–81 (2014)
23. Manly, T.: The absent mind: further investigations of sustained attention to response. Neuropsychologia **37**(6), 661–670 (1999)
24. OculusRift-Virtual Reality Headset for 3D Gaming Homepage. https://www.oculus.com/. Accessed 19 Jan 2019
25. Pasini, M., Berto, R., Brondino, M., Hall, R., Ortner, C.: How to measure the restorative quality of environments: the PRS-11. Procedia - Soc. Behav. Sci. **159**, 293–297 (2014)

26. Mueller, S.T., Piper, B.J.: The psychology experiment building language (PEBL) and PEBL test battery. J. Neurosci. Methods **222**, 250–259 (2014)
27. R Core Team: R: A language and environment for statistical computing. R Foundation for Statistical Computing, Vienna, Austria (2012). http://www.R-project.org/. ISBN 3-900051-07-0
28. Aiken, L.S., West, S.G.: Multiple Regression: Testing and Interpreting Interactions. Sage, Thousand Oaks (1991)
29. Berman, M.G., Jonides, J., Kaplan, S.: The cognitive benefits of interacting with nature. Psychol. Sci. **19**(12), 1207–1212 (2008)
30. Berman, M.G., Kross, E., Krpan, K.M., Askren, M.K., Burson, A., Deldin, P.J., Kaplan, S., Sherdell, L., Gotlib, I.H., Jonides, J.: Interacting with nature improves cognition and affect for individuals with depression. J. Affect. Disord. **140**(3), 300–305 (2012)
31. Bronack, S., Sanders, R., Cheney, A., Riedl, R., Tashner, J., Matzen, N.: Presence pedagogy: teaching and learning in a 3D virtual immersive world. Int. J. Teach. Learn. High. Educ. **20**(1), 59–69 (2008)
32. Riva, G.: Virtual reality: an experiential tool for clinical psychology. Br. J. Guid. Couns. **37**(3), 337–345 (2009)
33. Li, H., Daugherty, T., Biocca, F.: Characteristics of virtual experience in electronic commerce: a protocol analysis. J. Interact. Mark. **15**(3), 13–30 (2001)
34. Freeman, D., Reeve, S., Robinson, A., Ehlers, A., Clark, D., Spanlang, B., Slater, M.: Virtual reality in the assessment, understanding, and treatment of mental health disorders. Psychiatry Med. **47**(14), 2393–2400 (2017)'

Technology-Based Trainings on Emotions: A Web Application on Earthquake-Related Emotional Prevention with Children

Daniela Raccanello(✉), Giada Vicentini, Margherita Brondino, and Roberto Burro

Department of Human Sciences, University of Verona, Verona, Italy
{daniela.raccanello,giada.vicentini,
margherita.brondino,roberto.burro}@univr.it

Abstract. In light of their potential for learning and engagement, using technology-based programs can be particularly relevant to enhance children's emotional competence, also in relation to traumatic events such as disasters. Some studies investigated the efficacy of technology-based interventions fostering this ability, focusing on its different components, with different populations, and using different designs, but they did not relate specifically to disasters such as earthquakes. Nevertheless, in everyday life knowledge on earthquakes can be promoted through the use of mobile applications. We searched electronically all the applications present within the Google Play Store, identifying 20 applications on earthquake prevention. None of them was specifically focused on earthquake-related emotional contents, but some of them included some emotional elements. Therefore, to fill in the gaps in the current psychological literature, we developed a web application to promote earthquake-related emotional knowledge, to be tested in the future according to the standards of evidence-based research.

Keywords: Evidence-based interventions · Mobile-App · Emotions · Earthquakes

1 Introduction

Current methodologies for the design of accessible and usable technology-enhanced learning (TEL) systems make it possible to develop learning trainings with educational aims, also focusing on socio-emotional learning. However, emotional trainings on disaster-related emotions, and specifically on earthquakes, are rare. Therefore, within a larger project (PrEmT project, Emotional Prevention and Earthquakes in primary school, http://www.dsu.univr.it/?ent=progetto&id=5125 we developed a web application promoting earthquake-related emotional competence, on the basis of the psychological state-of-the-art. The main aim of this web application is to increase knowledge on emotion understanding and emotion regulation. According to a psychological perspective [4], knowledge on emotions is a prerequisite for their efficacious use in everyday life, and these general mechanisms can be generalized also for the specific case of earthquake-related psychological processes.

R. Gennari et al. (Eds.): MIS4TEL 2019, AISC 1007, pp. 53–61, 2020.
https://doi.org/10.1007/978-3-030-23990-9_7

2 Use of Mobile Devices to Foster Emotional Competence

Within the field of learning and education, the use of technology through mobile devices (such as smartphones, tablets, and laptops) and desktop computers poses a great challenge both for learners and professionals developing trainings [9, 12]. According to a recent meta-analysis, the use of mobile devices within formal and informal learning contexts have increased significantly in the last years, and it is more effective for learning achievement than using traditional teaching methods such as pencil-and-paper but also desktop computers [22]. In addition, the educational programs based on mobile devices have the potential to sustain and promote students' motivation, engagement, autonomy, problem-solving, and creativity [9, 25]. In light of their potential for learning and engagement, using technology-based programs can be particularly relevant to enhance emotional competence, as the ability of expressing, understanding, and regulating emotions [4]. Such a skill represents a basic psychological resource and it is acquired gradually by children and adolescents along development [4].

Some studies investigated the efficacy of technology-based interventions with different methodologies aimed at fostering emotional competence, focusing on different components of emotional competence, populations, and designs. Overall, these studies used a variety of technology-based methodologies combining different hardware, such as computer-devices [5, 10, 19] and tablets [2], with many types of learning-oriented software, such as videogames [1, 6, 14], augmented or virtual reality programs [2, 3, 5], and other ad-hoc software [11, 15]. Emotional competence has been operationalized variously, according to its different components. The abilities of expressing and understanding emotions were trained through interventions mainly focused on prosodic aspects [1], facial expressions [3, 20, 23], and emotional concepts [11]. Other studies are aimed at teaching how to regulate emotions in order to enhance positive emotions and diminish negative ones [6, 7, 21, 24]. Furthermore, different studies involved different populations, such as non-clinical children [6, 7, 24], and children or adolescents with autism spectrum disorders [13], physical impairments [11], or other types of psychological or behavioural problems [3, 10, 14]. Concerning design type, only rarely these studies followed all the standards of evidence-based research, according to which an efficacious intervention should "be tested in at least two rigorous trials that (1) involved defined samples from defined populations, (2) used psychometrically sound measures and data collection procedures; (3) analysed their data with rigorous statistical approaches; (4) showed consistent positive effects [...]; and (5) reported at least one significant long-term follow-up" [8, p. 151]. In addition, many of these studies had a low number of participants; they frequently involved children with atypical rather than typical development; and none of them focused specifically on emotional competence related to disasters, and in particular earthquakes. Nevertheless, in everyday life knowledge on earthquakes can be promoted through the use of mobile applications.

3 Applications on Earthquakes

Although there are no applications focused on earthquake-related emotional competence, some mobile applications regard earthquake prevention in general. Particularly, it is interesting the case of those applications optimized for android technology, since it is the most widespread operating system for mobile-devices (http://gs.statcounter.com/os-market-share/mobile/worldwide).

3.1 Search Strategy for Applications on Earthquakes

We searched electronically all the applications present within the Google Play Store between 1st and 31st January 2019, using as keywords the English words "earthquake/s", and the Italian corresponding translations "terremoto/i". The initial search yielded more than 350 applications, including games, didactic applications, earthquakes trackers, and audio applications. Concerning inclusion and exclusion criteria, applications were selected if they conformed with the following three characteristics: (a) The name of the application focused on earthquake characteristics and/or safety behaviours; (b) English or Italian language; (c) and the application was not focused on alert messages on earthquakes occurrence. Using these criteria yielded 20 applications that were acceptable for this review. We report in Table 1 a short description of the characteristics of the applications corresponding to our selection criteria.

3.2 Characteristics of Applications on Earthquakes

The selected applications can be distinguished on the basis of two different teaching methods, i.e. *game-based* learning vs. *lecture-based* learning [e.g., 22]. In both cases, the applications had the aim to enhance people's knowledge on earthquakes, focusing mainly on the use of interactive games for game-based learning, and presenting written information, oral contents, or sounds typical of earthquakes for lecture-based learning.

The applications focused on game-based learning aim at teaching children and/or adults knowledge on earthquakes while having fun. Some applications are devoted to young children, with simple goals and oral instructions (A, B, C). On the other hand, games for older children or adults are usually characterized by a higher level of structuration and are based on written instructions and descriptions (D, E, F, G). On the whole, these applications include characters such as humans, animals (e.g., pandas), or animated objects (e.g., life ring), that have to save themselves or others. Using these applications, the player can learn how to behave during (e.g., sheltering under a table) or after an earthquake (e.g., listening to teachers or rescuers' indications) in different settings (e.g., open areas, schools, homes, or supermarkets), and which are the characteristics of earthquakes. The main aspects of the earthquakes are presented through images (e.g., depicting falling objects), audios (e.g., collapse noise), and video effects (e.g., vibration of the screen). Only rarely, fantasy elements such as dragons as causes of earthquakes are inserted, as in Tanah (G). Some of these games have also the aim to teach how to behave in other emergency situations, such as tsunamis (G), floods (E), or fires (F). The applications focused on lecture-based learning have explicit didactic aims. We can distinguish applications suitable for children, with a major level of

Table 1. Characteristics of selected applications on earthquakes: type (1 = Game-based applications; 2 = Lecture-based applications), name, language (E = English, I = Italian), developer, version, and last updating. In the text, the applications are referred to using the corresponding letter.

Type	Name	Language	Developer	Version	Last updating
1	A. Little Panda Earthquake Safety	E	BabyBus Kids Games	8.30.10.00	28/12/18
	B. Earthquake Safety Tips 2	E	BabyBus Kids Games	8.30.10.00	03/01/19
	C. Little Panda's Earthquake Rescue	E	BabyBus Kids Games	8.30.10.00	28/12/18
	D. Il Terremoto–Capitan Ciambella 2	I	BerGAME	1.1.15	28/10/18
	E. HELP-ME	I, E	Format Formazione Tecnica	1.0.0	30/10/17
	F. Hazard Rush	E	MGames ICTD	1.0	20/04/17
	G. Tanah	E	Opendream	1.1.6	20/12/16
2	H. Participating in Earthquake Drill	E	DOST–Science Education Inst.	1.0.0	05/12/15
	I. Analyzing Earthquake Hazards and Coping Mechanisms	E	DOST–Science Education Inst.	0.0.1	17/11/18
	J. Earthquake	E	DOST–Science Education Inst.	1.0.0	22/01/19
	K. Is That an Earthquake? Ready to Read	E	Ministry of Education, New Zealand	2	05/10/14
	L. Earthquake Survival Tips	E	Cooler	2.3	23/08/15
	M. Earthquake	E	Let's Learn Something	1.0	19/11/18
	N. Emergency preparedness & Disaster Survival Guide	E	Summer Rabbit	1.0.6	05/11/18
	O. Earthquake Sounds Ringtones	E	BCS Developer	2.6.9	10/09/18
	P. Terrifying earthquake sounds	E	Corrado Borgatti	1.1	01/02/15
	Q. Earthquake Sound	E	Infinite_Apps	1.41	23/10/18
	R. Earthquake Audio Ringtone	E	Junscloud Developer	2.6.9	16/09/18
	S. Earthquake Sounds Ringtone	E	Sugar Daddy Developer	9.0	13/04/18
	T. Appp.io - Disaster Sounds	E	Appp.io	1.0.2	17/10/18

interactions and the addition of images and sound effects (H, I, J, K), and applications more appropriate for adults, with more written information and less animations (L, M, N). These didactic applications are more focused on earthquake-related contents than game-based applications. The user can learn which are the typical hazards and causes of an earthquake and how to behave before, during, and after it. In those applications designed for children, the use of images, sounds, and video effects facilitates learning. Among them, one focuses also on other types of natural disasters, such as tsunamis, tornados, and snowstorms (N). A sub-category of lecture-based learning applications includes sound applications, through which it is possible to download and/or listen to ringtones about earthquakes (O, P, Q, R, S) and other natural disasters (T). These types of applications are interesting because they can give examples of typical earthquake-related sounds, such as loud and collapse noises, enabling people to experience directly some of the key characteristics of this phenomenon. Some of these applications contain also a set of safety behaviours to be implemented (P, R).

3.3 Applications Including Earthquake-Related Emotional Elements

None of the selected applications was specifically focused on earthquake-related emotional contents. However, some of them included incidentally some elements pertaining to emotional competence. Concerning expression of emotions, some applications include characters showing negative emotions through facial expressions. There are baby rabbits, wolves, and mice crying desperately; there are scared baby pandas; there is a worried teacher represented by a sheep (A, B, C). There are also scary boys and girls, children trembling, or list of words describing emotions (E, G, K). On the whole, most of the expressed emotions refers to fear/anxiety or sadness; however, also a larger variety of emotions is expressed, such as calm, surprise, relieve (E), confirming previous literature and data on both children and adults referring to the emotions that can be felt during and just after an earthquake [16–18]. As regards understanding and regulation of emotions, the selected applications did not include tasks aiming at practicing the ability of emotion recognition from facial expressions or using psychological lexicon to describe emotions. However, whether specific tasks are inserted, they concern mostly emotion regulation. The main conveyed message focuses on the relevance of staying calm and not panicking (A, D, G, H, N). Such strategies can be classified as pertaining to self-reliance coping, according to which people cope with stressful events focusing on personal emotional resources [16, 18, 26]. Other adaptive regulation strategies include social sharing and support, highlighting the relieving benefits of talking about earthquake-related damage and feelings, as in Tanah (G). In most of the cases, the applications include information on adaptive and maladaptive regulation strategies without giving to the users the possibility to interact. Usually, within game-based applications such information is exchanged between characters, for example in dialogues (A, G); within lecture-based applications, it is presented as safety tips (N). Exceptions are applications such as Little Panda Earthquake Safety Tips or HELP-ME, in which users can choose the appropriate reaction (A, E). In the former, final forced-choice questions ask the user to identify the properness of reactions such as desperate crying during the shakes of an earthquake. In the latter, users are presented short stories followed by forced-choice questions, presenting alternatives such as behaving according to safety indications vs. ignoring them for the overwhelming effects of panic.

4 A Web Application on Earthquake-Related Emotional Prevention with Children

To fill in the gaps in the current psychological literature, we have developed a web application to promote earthquake-related knowledge, focusing on emotional competence. The web application is called E:READ-Y, Earthquakes: Ready for Emotions Associated with Disasters-Young version. As part of a larger project, i.e. the PrEmT project, the web application will be tested within an evidence-based intervention with primary school children, and then adjusted to be used for both children and adults. The project is interdisciplinary in nature, and it is conducted with the collaboration of professionals from the fields of psychology, geology, education, informatics, and illustration for children.

4.1 Aim of E:READ-Y

E:READ-Y is addressed to both children and adults to promote the development of knowledge on earthquakes. Preliminary knowledge regards characteristics of earthquakes and safety behaviours, and key knowledge focuses on general and earthquake-related emotions and coping strategies.

4.2 Characteristics of E:READ-Y

Differently from most of the cited existing applications, E:READ-Y is both game-based and lecture-based; in addition, it is a web application, with related advantages. It includes ten levels. The first two preliminary levels focus on geological contents, i.e. characteristics of earthquakes and safety behaviours, while the other eight units focus on psychological emotional contents. Each level includes a number of items from about 24 to 48, in which users have to respond to dichotomous forced-choice questions. Each item includes a stimulus formed by sentences and/or images and/or sounds. The passage from one level to the following is possible when the user reaches a minimum score.

As follows, the main technical characteristics of the web application are described.

The web-application is a server-side dynamic web software whose management is controlled by Microsoft Internet Information Server (IIS) application processing server-side scripts written for ASP.NET Core (a cross-platform, high-performance, open-source framework for building performing modern, cloud-based, and internet-connected applications). In order to store and retrieve data, a MS-sql relational database management system is used. A Responsive Web Design (RWD) is adopted, namely the web page interface on an internet browser fits well for a variety of devices and window or screen sizes (personal computers, notebooks, tablets, or smartphones).

The web application has a front-office and a back-office structure. The language of the web application is Italian; the development of the English version is ongoing.

5 Conclusions

We described the development of a web application aimed at fostering children's knowledge on earthquakes, focusing on emotional competence. As regards the scientific impact of this paper, examining the current psychological literature concerning technology-based programs on emotional competence and existing applications on earthquake-related emotions enabled us to map the current state-of-the-art on this issue, neglected by the current literature. Concerning the technological impact, we used Information and Communication Technology (ICT) for developing technological instruments ad-hoc, e.g. the E:READ-Y web application, to be included in an educational training specifically addressed to children. Finally, we highlight the social impact of the whole PrEmT project in terms of emotional prevention, contributing to reduce the cost of psychological post-traumatic interventions for managing emotional negative consequences related to traumatic events such as earthquakes.

This paper suffers from some limitations. First, the review of the earthquake-related applications was limited to the android technology; future studies should include also different technologies. Second, we excluded earthquake trackers applications; future studies should also explore the contents of these applications, to search for contents on earthquakes-related characteristics, safety behaviours, and emotional elements. Third, the description of E:READ-Y focused on its technical characteristics, but we did not report data on its usability, that will be gathered in the future. Notwithstanding these limitations, our proposal is a first step to develop a web application on earthquake-related emotional competence for children. Its efficacy will be tested according to the standards of evidence-based research [8], to help improving current methodologies for the design of accessible and usable TEL systems.

References

1. Brooks, P.J., Ploog, B.O.: Attention to emotional tone of voice in speech perception in children with autism. Res. Autism Spectr. Disord. **7**(7), 845–857 (2013). https://doi.org/10.1016/j.rasd.2013.03.003
2. Chen, C.H., Lee, I.J., Lin, L.Y.: Augmented reality-based video-modeling storybook of nonverbal facial cues for children with autism spectrum disorder to improve their perceptions and judgments of facial expressions and emotions. Comput. Hum. Behav. **55**(part A), 477–485 (2016). https://doi.org/10.1016/j.chb.2015.09.033
3. Cheng, Y., Chen, S.: Improving social understanding of individuals of intellectual and developmental disabilities through a 3D-facial expression intervention program. Res. Dev. Disabil. **31**(6), 1434–1442 (2010). https://doi.org/10.1016/j.ridd.2010.06.015
4. Denham, S.A.: Emotional Development in Young Children. Guilford, New York (1998)
5. Didehbani, N., Allen, T., Kandalaft, M., Krawczyk, D., Chapman, S.: Virtual reality social cognition training for children with high functioning autism. Comput. Hum. Behav. **62**, 703–711 (2016). https://doi.org/10.1016/j.chb.2016.04.033

6. Filella, G., Cabello, E., Pérez-Escoda, N., Ros-Morente, A.: Evaluation of the emotional education program "Happy 8-12" for the assertive resolution of conflicts among peers. Electron. J. Res. Educ. Psychol. **14**(3), 582–601 (2016). https://doi.org/10.14204/ejrep.40.15164
7. Filella, G., Ros-Morente, A., Oriol, X., March-Llanes, J.: The assertive resolution of conflicts in school with a gamified emotion education program. Front. Psychol. **9**, 2353 (2018). https://doi.org/10.3389/fpsyg.2018.02353
8. Flay, B.R., Biglan, A., Boruch, R.F., Ganzalez Castro, F., Gottfredson, D., Kellam, S., et al.: Standards of evidence: criteria for efficacy, effectiveness and dissemination. Prev. Sci. **6**(3), 151–175 (2005). https://doi.org/10.1007/s11121-005-5553-y
9. Fleischer, H.: What is our current understanding of one-to-one computer projects: a systematic narrative research review. Educ. Res. Rev. **7**, 107–122 (2012). https://doi.org/10.1016/j.edurev.2011.11.004
10. Glaser, B., Lothe, A., Chabloz, M., Dukes, D., Pasca, C., Redoute, J., Eliez, S.: Candidate socioemotional remediation program for individuals with intellectual disability. Am. J. Intellect. Dev. Disabil. **117**(5), 368–383 (2012). https://doi.org/10.1352/1944-7558-117.5.368
11. Goker, H., Ozaydin, L., Tekedere, H.: The effectiveness and usability of the educational software on concept education for young children with impaired hearing. Eurasia J. Math. Sci. Technol. Educ. **12**(1), 109–124 (2016). https://doi.org/10.12973/eurasia.2016.1207a
12. Graesser, A.C., D'Mello, S.K., Strain, A.C.: Emotions in advanced learning technologies. In: Pekrun, R., Linnenbrink-Garcia, L. (eds.) International Handbook of Emotions in Education, pp. 473–493. Taylor & Francis, New York (2014)
13. Grynszpan, O., Weiss, P.L., Perez-Diaz, F., Gal, E.: Innovative technology-based interventions for autism spectrum disorders: a meta-analysis. Autism **18**(4), 346–361 (2014). https://doi.org/10.1177/1362361313476767
14. Hobbs, L.J., Yan, Z.: Cracking the walnut: using a computer game to impact cognition, emotion, and behavior of highly aggressive fifth grade students. Comput. Hum. Behav. **24**(2), 421–438 (2008). https://doi.org/10.1016/j.chb.2007.01.031
15. Matsuda, S., Yamamoto, J.: Computer-based intervention for inferring facial expressions from the socio-emotional context in two children with autism spectrum disorders. Res. Autism Spectr. Disord. **8**(8), 944–950 (2014). https://doi.org/10.1016/j.rasd.2014.04.010
16. Raccanello, D., Barnaba, V., Rocca, E., Hall, R., Burro, R.: Children and adults' representation of emotions and coping strategies related to earthquakes (2019)
17. Raccanello, D., Burro, R., Hall, R.: Children's emotional experience two years after an earthquake: an exploration of knowledge of earthquakes and associated emotions. PLoS ONE **12**(12), 1–21 (2017). https://doi.org/10.1371/journal.pone.0189633
18. Raccanello, D., Rocca, E., Barnaba, V., Hall, R., Brondino, M.: Strategies to cope with natural disasters: A meta-analysis on children and adolescents (2018)
19. Romero, N.L.: A pilot study examining a computer-based intervention to improve recognition and understanding of emotions in young children with communication and social deficits. Res. Dev. Disabil. **65**, 35–45 (2017). https://doi.org/10.1016/j.ridd.2017.04.007
20. Russo-Ponsaran, N.M., Evans-Smith, B., Johnson, J.K., McKown, C.: A pilot study assessing the feasibility of a facial emotion training paradigm for school-age children with autism spectrum disorders. J. Ment. Health Res. Intellect. Disabil. **7**, 169–190 (2014). https://doi.org/10.1080/19315864.2013.793440
21. Schuurmans, A.A., Nijhof, K.S., Vermaes, I.P., Engels, R.C., Granic, I.: A pilot study evaluating "Dojo", a videogame intervention for youths with externalizing and anxiety problems. Games Health J. **4**(5), 401–408 (2015). https://doi.org/10.1089/g4h.2014.0138

22. Sung, Y.T., Chang, K.E., Liu, T.C.: The effects of integrating mobile devices with teaching and learning on students' learning performance: a meta-analysis and research synthesis. Comput. Educ. **94**, 252–275 (2016). https://doi.org/10.1016/j.compedu.2015.11.008
23. Thomeer, M.L., Rodgers, J.D., Lopata, C., McDonald, C.A., Volker, M.A., Toomey, J.A., et al.: Open-trial pilot of mind reading and in vivo rehearsal for children with HFASD. Focus. Autism Other Dev. Disabil. **26**(3), 153–161 (2011). https://doi.org/10.1177/1088357611414876
24. Verkijika, S.F., De Wet, L.: Using a brain-computer interface (BCI) in reducing math anxiety: evidence from South Africa. Comput. Educ. **81**, 113–122 (2015). https://doi.org/10.1016/j.compedu.2014.10.002
25. Warschauer, M.: A teacher's place in the digital divide. Yearb. Natl. Soc. Study Educ. **106**, 147–166 (2007). https://doi.org/10.1111/j.1744-7984.2007.00118.x
26. Zimmer-Gembeck, M.J., Skinner, E.A.: The development of coping across childhood and adolescence: an integrative review and critique of research. Int. J. Behav. Dev. **35**(1), 1–17 (2011). https://doi.org/10.1177/0165025410384923

Using Rasch Models for Developing Fast Technology Enhanced Learning Solutions: An Example with Emojis

Roberto Burro, Margherita Pasini, and Daniela Raccanello(✉)

Department of Human Sciences, University of Verona, Verona, Italy
{roberto.burro,margherita.pasini,
daniela.raccanello}@univr.it

Abstract. We focus on issues related to learning analytics for predicting behavior, presenting a case in which people's answers can be predicted on the bases of a known numeric function describing a cognitive relation. We used the Rasch scaling method following an item-response latent-trait model on questions on the valence of emojis of common use. Emojis are pictorial symbols with a high degree of humanization, nowadays increasingly frequent in computer-mediated communication (CMC). However, sometimes their meaning is ambiguous. Therefore, after quantifying how much of a property such as positive and negative valence different emojis represented, we identified the function that relates the two different judgements on the same object, specifically the positive and negative valence of each emoji. Applications of this approach can be highly relevant for learning analytics, to optimize the measurement, collection, and analysis of data about learners and their contexts for fast Technology Enhanced Learning (TEL) solutions.

Keywords: Rasch models · Psychophysical scaling · Emotions · Emojis · Learning · Fundamental measurement

1 Introduction

In this paper we focus on issues related to learning analytics for predicting behavior, presenting a case in which people's answers can be predicted on the bases of a known (i.e., previously calculated) numeric function that describes a cognitive relation. In other words, answers given to a set of questions enable to predict the answers to another related set of questions, which are however not administered, with an advantageous sparing of time. In our case, the set of questions focuses on the assessment of positive and negative valence of emojis. Applications of this approach can be highly relevant for learning analytics, in order to optimize the measurement, collection, and analysis of data about learners and their contexts for fast Technology Enhanced Learning (TEL) solutions.

R. Gennari et al. (Eds.): MIS4TEL 2019, AISC 1007, pp. 62–70, 2020.
https://doi.org/10.1007/978-3-030-23990-9_8

1.1 Properties and Functional Relations

Entities have many properties: For example, a person has a body-height, a body-weight, a risk propensity, an Intelligence Quotient (IQ), a level of extraversion, a certain kindness, etc. Some of these properties are related among them: Height and weight are clearly in direct relation (i.e., a tall person will presumably weigh more than a small person). Formally:

$$\begin{aligned} \mathrm{y} &= \mathrm{f(x)} \\ \mathrm{weight} &= \mathrm{f(height)} \end{aligned}$$

However, other properties are not related: For example, there is no relation between height, IQ, and extraversion. Knowing the type of functional relation existing between two properties, it is possible to estimate the value of the other beginning from only one of the two. The more precise is the function that describes the relation, the better is the estimate of the property that is not given. Moreover, it is possible to estimate the relation between more than two properties. It follows that it is useful to translate a function into an IT-algorithm (Information Technology-algorithm) that allows, through a systematic calculation procedure performed by a computer, to improve the amount of available information.

Let's take the example of the above mentioned height-weight relation. There exist different algorithms (e.g., Peterson formula, Robinson formula, Devine formula, Broca formula, Hamwi formula, Miller formula, Hammond formula, etc.) that allow to estimate the body-weight of a person starting from his/her body-height (in some cases considering also sex and age), gaining information initially not given (the weight) [22]. This leads to an advantageous logical consideration: The IT implementation of a function that describes the relation between properties allows a rapid gain of information. This can be useful in all the cases in which there is the need to gather, through the use of web-browsers or ad-hoc software, data for profiling, categorization, and/or evaluation purposes without spending too much time. Also in the psychological field, this happens often, for example administering questionnaires. The issue becomes even more evident in within-subject designs in which one person answers to a set of questions with reference, for example, to different contexts (e.g., real or imagined, referred to learning or job contexts, etc.).

1.2 How to Get a Function in Cognitive Contexts

Generally, there are two necessary elements to compute and implement a functional relation between two or more properties. We need: (1) statistical units on which it is possible to detect the properties of interest (i.e., in the case of height and weight properties, the statistical units are people); (2) fundamental measurements [11] of the properties (i.e., a measure in centimeters for height and a measure in kilograms for weight).

The first requirement can be found easily; however, the second is more difficult to be gained. While for disciplines like Physics, Engineering, Biology, etc., it is quite easy to have measuring instruments that quantify properties (i.e., meters, scales, etc.), it becomes difficult to have fundamental measures for all those cognitive properties that do not have a precise and shared measurement tools (i.e., predisposition to the risk, kindness, emotionality, etc.). However, in the cognitive contexts these constraints can be overcome through the use of the Rasch models [2, 27].

The Rasch scaling method follows an item-response latent-trait model. Starting from a set of survey items, it produces a unidimensional logit-interval scale relating to the extent to which each item conveys a latent property [6, 30]. The scale is then used to measure the participants' responses to the items. In this way, it is possible "to weight" the relative impact of the various items (i.e., the item content) and to investigate the effects on people.

Rasch models review the weak points of traditional approaches to measurement (the classic test theory) by giving priority to objective measurements of latent dimensions which are based on the principles of fundamental measurement [17]. In other words, individuals are measured independently of the characteristics of the items and items' calibrations are independent from the characteristics of the individuals. Fundamental measurement is taken for granted in the physical sciences, whereas in the social sciences the raw scores and the sum or means of these scores are typically considered as "measures" of a dimension independently from whether or not they conform to the principles of fundamental measurement [12].

The measures obtained by means of the application of the Rasch models have the following measurement properties: (a) linearity: The total score (i.e., sum, mean, etc.) relating to the raw scores given by the participants using a categorical or rating response scale do not constitute an interval measurement scale due to the non-linear metric nature of this procedure [3]; On the contrary, the unit of measurement (logit) defined by means of the application of Rasch models (logit) and derived from the logarithmic form of the model has the same constant value all along the continuum of the latent dimension; (b) specific objectivity: The relation between stimulus parameters defined by the application of a Rasch model is not influenced by subject parameters and vice versa; (c) stochastic independence: The probability associated with a pattern of responses given by a subject n to the stimulus i is the product of the response probabilities given to each of the i stimuli.

It is worth noting that the Rasch linear measures approach differs from other typical approaches based on raw scores [30] for a number of issues (Table 1).

1.3 Emojis: Definition, Functions, and Potential for Measurement in Learning Contexts

Emojis are pictorial symbols with a high degree of humanization [1]. They were derived from emoticons, i.e., pictorial representations of human facial expressions introduced during the Eighties as formed uniquely by standard keyboard typographical characters [1]. Nowadays, their use in computer-mediated communication (CMC) is constantly increasing [14]. From a psychological perspective, the function of emojis is the same for CMC of what nonverbal cues related to facial expressions, gestures, or tone of voice contributing to express emotions are for face-to-face communication [1]. From the perspective of the media richness theory, the combination of multiple information through

Table 1. Differences between the Rasch linear measures approach and approaches based on raw scores

Score characteristics	Other raw scores approaches	Our linear measures approach
Psychometrics theory	Classical test theory	Rasch model
Type	Ordinal ranking on hoped for latent variable	Linear positioning on latent variable explicitly defined by item content
Structure	Non-linear (no additivity)	Linear (straight additivity)
Missing data	The score is highly distorted by the presence of missing data and un-administered items	The score is not altered by missing data and un-administered items
Veracity	Mistaken for truth	Known because estimated
Continuity	Not continuous, but discrete	Continuous
Precision	Unknown, except on average	Quantified by standard errors
Accuracy	Unknown	Quantified by fit statistics
Parametric-linear analysis	Unsuited	Ideal
Validity	Sample-dependent test reliability	Item-dependent construct validity
Reproducibility	Irreproducible	Reproducible
Generalizability	Local	General
Comparability with other scores	Hard	Easy
Dependence	Test-bound	Test-free
Integration	Troublesome	Easy

texts and emojis would increase richness in communication, facilitating effective CMC [15]. However, even if the meaning of emojis is quite intuitive and also published in reference websites as Emojipedia [13], sometimes emojis result ambiguous.

Identifying precisely the meaning of emojis could be highly relevant for assessment issue, given that they could be used as stimuli included in self or other-report instruments to assess the emotions associated to a variety of constructs, also for TEL contexts. Indeed, notwithstanding the possible ambiguity in their meanings, emojis are already amply used in the evaluation of a variety of objects in technological environments, to evaluate the valence or to gather opinions on products, services, issues, policies, and health-care [3, 20].

1.4 Aims of This Study

Our main aim was to identify the function that relates two different judgements on a same object, specifically the positive and negative valence of an emoji. Specifically, we aimed at: (a) quantifying how much of a property such as positive and negative valence each emoji represents; (b) investigating whether the two properties are significantly related; and (c) describing this relation in order to identify a function to be used to make predictions.

2 Method

2.1 Participants

We involved 110 undergraduate students (M = 23.24 years, SD = 6.40; 37% male) at the University of Verona, in Northern Italy, with varying socio-economic status. The students participated after signing an informed consent form.

2.2 Material and Procedure

We gathered the data during 2018 through a questionnaire administered with mixed devices, i.e., smartphones or laptops. The questionnaire included measures on the valence of the presented emojis, and the data were preliminarily analysed in order to quantify the amount of positive and negative valence expressed by each of them [7]. We recruited the students during university lessons, and we administered the questionnaire through the Cognitive Metrix Survey Software, CMSS [8, 21, 23–26, 28]. We selected 81 emojis from the palette of emojis available in WhatsApp cross-platform instant messaging service during January 2018, according to the following criteria: (a) being drawn as stylized faces through a circle; (b) being of the same dimensions. The emojis were presented twice, in two blocks. For each emoji in each block, we asked to indicate how much positive and how much negative the expressed emotion was (*How much positive/negative is the expressed emotion?*). The participants could answer moving a slider along a bar ranging from 0 (*not at all*) to 100 (*very much*).We adapted the task from previous ones [14].

3 Results and Discussion

Preliminarily, we carried out the Rasch analysis using the Partial Credit Model [19] through the eRm-package of the R-software [18].

Through the application of Rasch models [2, 5, 9], we quantified the amount of positive and negative valence expressed by each emoji. To evaluate the goodness-of-fit of the emojis to the Rasch models (examining each emoji for both dimensions, i.e. positive and negative) we considered two criteria. Emojis have a good fit when the infit-t statistics relating to each of them is included between −2 and 2 [16, 29], and the p-value associated to the Chi square is higher than .05 [4]. We deleted the items with a bad fit to improve the measurement properties of the whole set of emojis. The scales were unidimensional, and there was not significant local dependency between items. Table 2 shows the 43 emojis who had a good fit for both dimensions.

Table 2. List of emojis with good fit and their scaling values for the positive and negative scale

Emoji	Positive value	Negative value	Emoji	Positive value	Negative value	Emoji	Positive value	Negative value
	0.001	2.223		1.697	1.071		3.482	0.587
	0.641	2.124		1.022	1.001		1.626	1.102
	1.266	1.932		1.029	1.222		2.150	0.871
	0.951	1.900		0.841	1.369		3.294	0.366
	0.568	1.854		0.944	1.404		3.072	0.464
	1.178	1.777		0.953	1.400		3.326	0.433
	0.951	1.632		2.066	1.120		3.209	0.426
	1.598	1.324		1.294	1.249		3.277	0.425
	1.111	1.400		1.308	1.243		3.197	0.386
	0.962	1.432		1.299	1.247		3.517	0.300
	0.823	1.237		3.125	0.440		3.544	0.400
	1.089	1.340		1.353	1.223		3.488	0.222
	0.789	1.289		2.290	0.632		3.541	0.001
	1.281	1.161		3.086	0.612			
	0.916	1.333		2.822	0.608			

Then, the set of 43 emojis that had a good fit for both dimensions was used to study the relations among the two dimensions. We conducted two models. In the first model (described in the first raw of Table 3), we predicted the negative scaling values from the positive scaling values; in the second model (described in the second raw of Table 3), we predicted the positive scaling values from negative scaling values.

Table 3. Linear models describing the relations between the positive and negative scaling values (Neg = negative scaling value; Pos = positive scaling value)

Function	Standard error	t value	Pr (> |t|)	R-squared
Neg = −0.475 × Pos + 1.948	0.031	−15.360	< .001	0.852
Pos = −1.793 × Neg + 3.769	0.117			

The linear function models turned out to have the better fit to our data (i.e., a simpler function and a higher R^2), compared to the fits of the logarithmic, power, and exponential functions. The significant relations among positive scaling values and negative scaling values of emojis are showed in Fig. 1 and reported in Table 3.

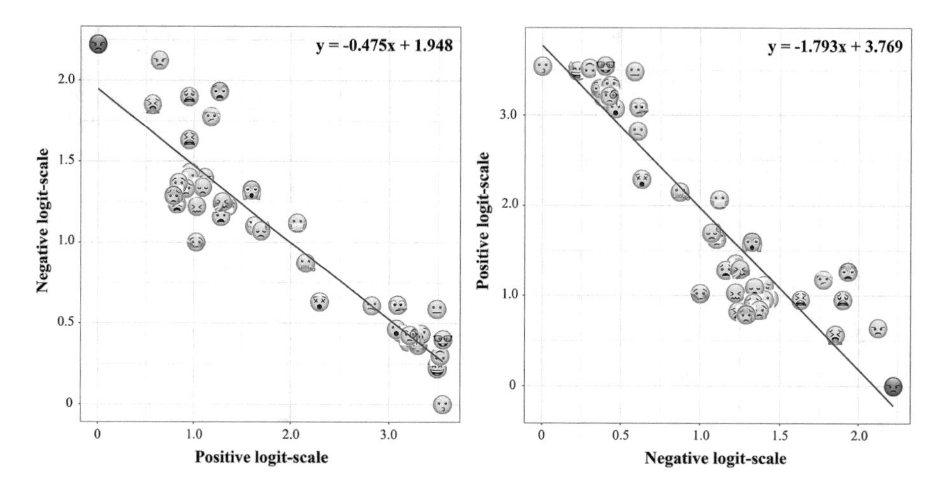

Fig. 1. Linear models describing the relations among the positive and negative scaling values, including the emojis. On the left, the horizontal axis represents the scaling values related to the assessment of positivity, while the vertical axis represents the scaling values related to the assessment of negativity, and vice versa on the right.

4 Conclusions

By means of the identified functions, it was possible to study a person's responses on one scale and to make a likely estimate of his/her responses on another scale [10], just like in the aforementioned formulas when we estimate the weight of people starting from their height. This gives a gain of information. In our case, when using the obtained formula we can ask to one person to assess the positivity of an emoji, and at the same time gather information on how much negative it is, and also vice versa. The potential of this approach for TEL contexts can be appreciated in terms of diminishing the effort for gathering psychological data, obtaining the same amount of information in a reduced time (half time in this specific case). Another benefit is the possibility to obtain also in TEL contexts psychological measures with properties which are qualitatively better of what is usually done with other approaches, as illustrated in Table 1.

References

1. Aldunate, N., González-Ibáñez, R.: An integrated review of emoticons in computer-mediated communication. Front. Psychol. **7i**(2061), 1–6 (2017). https://doi.org/10.3389/fpsyg.2016.02061
2. Andrich, D.: Rasch Models for Measurement. Quantitative applications in the social sciences, vol. 68. Sage, London (1988)
3. Asghar, M.Z., Khan, A., Ahmad, S., Qasim, M., Khan, I.A.: Lexicon-enhanced sentiment analysis framework rule-based classification scheme. PLoS ONE **12**(2), 1–22 (2017). https://doi.org/10.1371/journal.pone.0171649

4. Bland, J.M., Altman, D.G.: Multiple significance tests: the Bonferroni method. Br. Med. J. **310**(6973), 170 (1995)
5. Bond, T., Fox, C.M.: Applying the Rasch Model: Fundamental Measurement in the Human Sciences, 3rd edn. Routledge, New York (2015)
6. Burro, R.: To be objective in experimental phenomenology: a psychophysics application. Springerplus **5**(1), 1720 (2016). https://doi.org/10.1186/s40064-016-3418-4
7. Burro, R., Raccanello, D., Pasini, M.: Emojis' psychophysics: measuring emotions in technology enhanced learning contexts. Adv. Intell. Soft Comput. **804**, 70–78 (2019). https://doi.org/10.1007/978-3-319-98872-6_9
8. Burro, R., Raccanello, D., Pasini, M., Brondino, M.: An estimation of a nonlinear dynamic process using latent class extended mixed models: affect profiles after terrorist attacks. Nonlinear Dyn. Psychol. Life Sci. **22**(1), 35–52 (2018)
9. Burro, R., Sartori, R., Vidotto, G.: The method of constant stimuli with three rating categories and the use of Rasch models. Qual. Quant. **45**(1), 43–58 (2011). https://doi.org/10.1007/s11135-009-9282-3
10. Burro, R., Savardi, U., Annunziata, M.A., De Paoli, P., Bianchi, I.: The perceived severity of a disease and the impact of the vocabulary used to convey information: using Rasch scaling in a simulated oncological scenario. Patient Prefer. Adherence **12**, 2553–2573 (2018). https://doi.org/10.2147/PPA.S175957
11. Campbell, N.R.: Physics: The elements. Cambridge University Press, Cambridge (2013)
12. Cavanagh, R.F., Romanoski, J.T.: Rating scale instruments and measurement. Learn. Environ. Res. **9**(3), 273–289 (2007)
13. Emojipedia (2018). http://emojipedia.org
14. Gallo, K.E., Swaney-Stueve, M., Chambers, D.H.: A focus group approach to understanding food-related emotions with children using words and emojis. J. Sens. Stud. **32**(e12264), 1–10 (2016). https://doi.org/10.1111/joss.12264
15. Hsieh, S.H., Tseng, T.H.: Playfulness in mobile instant messaging: Examining the influence of emoticons and text messaging on social interaction. Comput. Hum. Behav. **69**, 405–414 (2017). https://doi.org/10.1016/j.chb.2016.12.052
16. Linacre, J.M.: What do infit and outfit, mean-square and standardized mean? Rasch Meas. Trans. **16**(2), 878 (2002)
17. Luce, R.D., Krantz, D.H., Suppes, P., Tversky, A.: Foundations of Measurement, vol. 3. Academic, San Diego (1990)
18. Mair, P., Hatzinger, R.: Extended Rasch modeling: the eRm package for the application of IRT models in R. J. Stat. Softw. **20**(9), 1–20 (2007)
19. Masters, G.N.: A Rasch model for partial credit scoring. Psychometrika **47**(2), 149–174 (1982)
20. Meschtscherjakov, A., Weiss, A., Scherndl, T.: Utilizing emoticons on mobile devices within ESM studies to measure emotions in the field. In: MobileHCI 2009: Proceedings of the 11th International Conference on Human-Computer Interaction with Mobile Devices and Services, pp. 1–4 (2009)
21. Pasini, M., Brondino, M., Burro, R., Raccanello, D., Gallo, S.: The use of different multiple devices for an ecological assessment in psychological research: an experience with a daily affect assessment. Adv. Intell. Soft Comput. **478**, 121–129 (2016). https://doi.org/10.1007/978-3-319-40165-2_13
22. Peterson, C.M., Thomas, D.M., Blackburn, G.L., Heymsfield, S.B.: Universal equation for estimating ideal body weight and body weight at any BMI. Am. J. Clin. Nutr. **103**, 1197–1203 (2016)

23. Raccanello, D., Brondino, M., Pasini, M., Landuzzi, M.G., Scarpanti, D., Vicentini, G., Massaro, M., Burro, R.: The usability of multiple devices for an ecological assessment in psychological research: salience of reasons underlying usability. Adv. Intell. Soft Comput. **804**, 79–87 (2019). https://doi.org/10.1007/978-3-319-98872-6_10
24. Raccanello, D., Burro, R., Brondino, M., Pasini, M.: Use of internet and wellbeing: a mixed-device survey. Adv. Intell. Soft Comput. **617**, 65–73 (2017). https://doi.org/10.1007/978-3-319-60819-8_8
25. Raccanello, D., Burro, R., Brondino, M., Pasini, M.: Relevance of terrorism for Italian students not directly exposed to it: the affective impact of the 2015 Paris and the 2016 Brussels attacks. Stress. Health **34**(2), 338–343 (2018). https://doi.org/10.1002/smi.2793
26. Raccanello, D., Burro, R., Hall, R.: Children's emotional experience two years after an earthquake: an exploration of knowledge of earthquakes and associated emotions. PLoS ONE **12**(2), 1–21 (2017). https://doi.org/10.1371/journal.pone.0189633
27. Rasch, G.: Probabilistic Models for Some Intelligence and Attainment Tests. University of Chicago Press, Chicago (1980)
28. Schmitz, C.: LimeSurvey: an open source survey tool. LimeSurvey Project Hamburg, Germany (2015). http://www.limesurvey.org
29. Wright, B.D., Linacre, J.M.: Reasonable mean-square fit values. Rasch Meas. Trans. **8**, 370–371 (1994)
30. Wright, B.D., Masters, G.N.: Rating Scale Analysis. Rasch Measurement. Mesa Press, Chicago (1982)

Immersive Virtual Reality in Technical Drawing of Engineering Degrees

M. P. Rubio[1(✉)], D. Vergara[2], and S. Rodríguez[1]

[1] University of Salamanca, Salamanca, Spain
{mprc, srg}@usal.es
[2] Catholic University of Ávila, Ávila, Spain
diego.vergara@ucavila.es

Abstract. The use of virtual reality (VR) technology is reaching almost all sectors: medicine, engineering, entertainment, defense, marketing, etc. Especially to those whose main objective is the representation of elements and three-dimensional environments of reality. This is especially relevant within the field of teaching-learning activities of technical drawing. In this case, one of the competences is to represent three-dimensional pieces and objects in two-dimensional drawings. In this process the student must have and develop the spatial skills that depend on the innate abilities of each one. In this communication an experience of the use of the Immersive Virtual Reality is presented to the development of spatial skills in technical drawing. This work describes the design, creation and programming of the computer application, choice of equipment and development of a methodology for use in teaching.

Keywords: Immersive virtual reality · Technical drawing · Spatial skills · Virtual environments

1 Introduction

Within the curricula of engineering and architecture degrees, the technical drawing assignments occupy a fundamental basis in the competences that students must acquire. The technical drawing is the graphic representation of an object or a practical idea that is guided by fixed and pre-established rules in order to accurately and clearly describe dimensions, forms and characteristics of what is wanted to be reproduced. In this process, spatial skills are very important [1, 2]. Defined as the ability to mentally manage complex three-dimensional shapes, it is of vital importance in the training stage and in the future professional life of the engineer [3, 4]. In addition, although the spatial ability is a personal skill that depends on the innate capacities of the individual (almost null in some people and very developed in others), it can be improved through training using methodologies that favor its development [5]. The methodologies that seem most effective are those based on the use of technology enhanced learning (TEL) tools. They model reality in the form of 3D graphics with a high degree of interaction that allow the simulation and experimentation of various phenomena or situations of reality [6–12].

This type of interactive three-dimensional graphic applications is generically called "Virtual Reality (VR)", since they are generated mathematically in a computer. The

R. Gennari et al. (Eds.): MIS4TEL 2019, AISC 1007, pp. 71–79, 2020.
https://doi.org/10.1007/978-3-030-23990-9_9

typology of three-dimensional virtual environments used in teaching-learning processes, ranging from the relatively simple to the very complex. But there are two general categories (Fig. 1) based display devices and interaction employees [13]: (i) non-immersive (window in the world), in which the user's vision to the world is through the flat computer screen (which acts as a window [14]) and (ii) the immersive ones, which completely introduce the user to the virtual world through glasses with two mini screens in front of the eyes and positioning systems that detect it in the real environment and place it in the virtual one. This communication describes the development of an immersive virtual reality tool applied in technical drawing, its design and its methodological implementation in the classes.

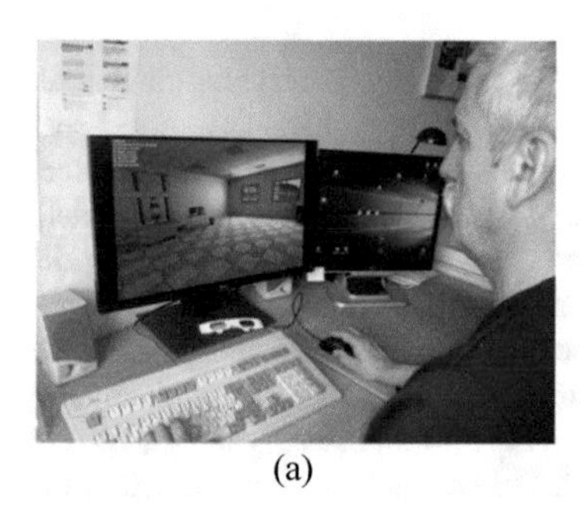

(a)

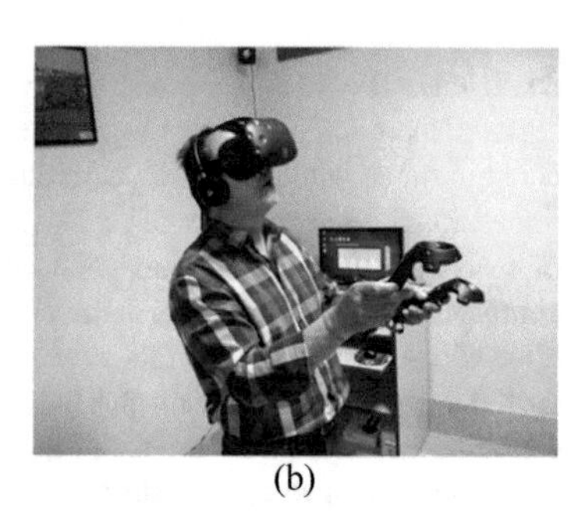

(b)

Fig. 1. Main types of VR systems [13]: (a) non-immersive; (b) immersive.

2 Immersive Virtual Reality (IVR)

From the point of view of current applications of virtual reality to engineering learning, most are simulations of three-dimensional virtual laboratories (3DVL) designed to perform practical sessions with complex, large, expensive or dangerous equipment that make difficult the use by of a large number of students [15, 16].

However, the use of virtual reality in engineering education extends beyond the use of 3DLV. On the one hand, virtual reality applications focus on the design and simulation of an engineering project, which are based not only on the use of techniques but also on the interactive verification of the results [17]. On the other hand, other applications of virtual reality aim to improve the understanding of different concepts: spatial understanding of abstract concepts, complex three-dimensional graphics, production processes, manufacturing, operation processes, assembly, etc. [18–21].

The most advanced uses in VR are focused on the simulation of a three-dimensional environment created by computer. In it, the user feels that he is in a virtual world that he perceives through different devices: glasses, controls, gloves, special suits, etc. [17, 22].

This type of VR is known as immersive virtual reality (IVR). The term immersive is added to make reference to the fact that the user can immerse himself in the virtual world, make movements and interact with virtual objects and people with an experience close to what he would experience in the real world.

The IVR is still an emerging technology and its full potential is not known in the education field, but it is known that it allows learning based on scenarios and experiences [23–25]. In addition, the use of IVR has other generic advantages [26]:

- *Improves comprehension*: Sometimes it is difficult to achieve the total comprehension of a concept, fundamentally when it is very abstract or complex. Therefore, showing it in all its dimensions with virtual reality can facilitate its understanding and correct processing.
- *Saves time*: As knowledge is understood more quickly, the results can be obtained in less time, so the teacher can spend more time in class explaining other concepts adjacent to the main information that he wants to transmit.
- *Generates more lasting knowledge*: What is experienced is remembered longer and better clarity than what someone else tells us. That is why the lessons that include an emotional part, such as those that can occur with VR, create more stable and lasting knowledge over time.
- *Encourages student attention and cooperation among them*: The creation of immersive experiences can increase student attention times as well as facilitate teamwork habits.

3 Implementation of the IVR in the Technical Drawing

The process of incorporating the IVR into the classes is complex and requires different tasks of design, decision making, choice of technologies and development of educational methodologies. In the following sections the tasks carried out in this project are summarized.

3.1 IVR Application Design

In this case, the developed application will be used in the subjects of normalization and representation of corporeal forms and three-dimensional volumes that require spatial skills. In these subjects, the use of solid three-dimensional pieces makes them ideal for digital modeling and presentation in three-dimensional immersive virtual reality environments.

Once the viability of applying the IVR to the selected topic is verified, the process of creating the application follows the steps shown in Fig. 2 [13]. In traditional engineering the pieces are designed by drawing their orthographic views (plant, elevation and profile) and with them the three-dimensional solid piece is manufactured. In the Reverse Engineering that is taught in the classes, the student starts from the piece in three dimensions drawn in a perspective system and has to obtain the orthographic views necessary for the correct definition of the piece (Fig. 3).

This method presents problems due to the deformations inherent to the perspective systems that can lead to errors in the visualization of the views and in the sizing of the pieces. In order to solve this problem, the perspective drawings can be replaced by real pieces (Fig. 4), but this requires having a large number of pieces for the practices and since they are different, not all the students in an evaluated exercise can use the same piece.

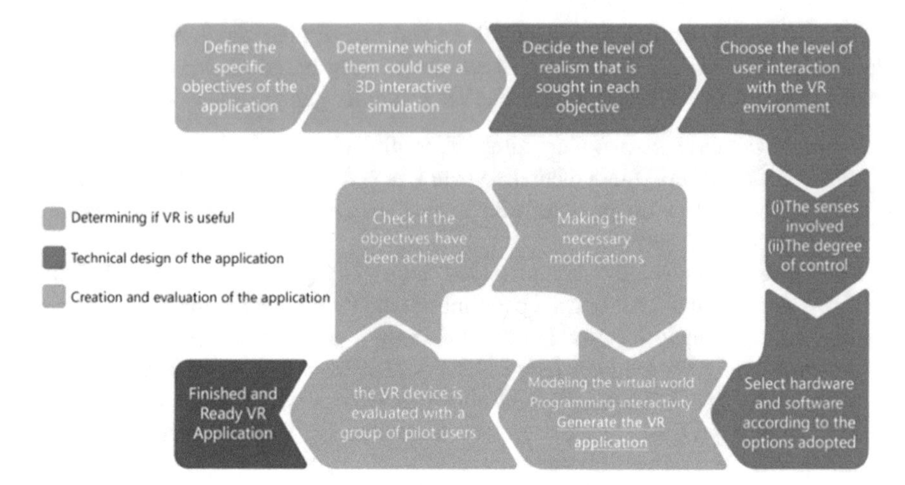

Fig. 2. Design scheme application IVR.

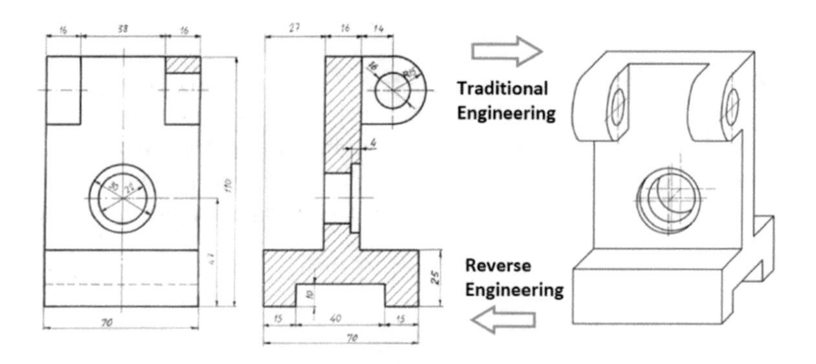

Fig. 3. Orthographic views and perspective of a piece.

Fig. 4. Collection of real pieces for technical drawing practical sessions.

In the practices with IVR, an interactive level has been chosen that allows exploring the environment taking advantage of the locomotion characteristics of the HTC Vive system and interacting with the parts with the controllers; taking them, rotating them and choosing the solutions of the corresponding views.

3.2 Choice of Hardware and Software

The development of the VR application requires a high-performance computer, with a powerful graphic card for use with the virtual reality system. The choice of the HTC Vive virtual reality system has been made following the "VR/AR Innovation Report" published by the Virtual Reality Developer Congress [27] and that designates this system as the most used and with the greatest future projection. This system is also fully compatible with the development engine chosen for the programming of the immersive virtual reality tool. The HTC Vive virtual reality system consists of glasses, controls and positioning bases and allows the user to walk around the environment.

Autodesk 3DS Max has been selected to perform the modeling of environments and parts, mainly due to the versatility and the large amount of documentation that exists in this software. It is the most used in engineering, architecture and in the creation of games. The programming of the virtual reality application has been carried out using Unreal Engine 4 (UE4) [28] of "Epic Games", which allows programming with a network system of interconnected nodes (called blueprint) that facilitate the work. This programming engine is used in the creation of commercial video games and its license is free if it is not used for profit.

3.3 Process of Creating the IVR Application

The pieces and the three-dimensional environments that contain them have been modeled using as reference real objects of the collections that have the teachers who teach the subjects (Fig. 4) and paying special attention to the details of form, materials and lighting. The three-dimensional environment has been chosen to be as simple as possible so that it does not interfere in the visualization of the pieces, although it must be realistic with the most appropriate scale, lighting, shadows and authentic materials. They are modeled with the Autodesk 3D Studio MAX program (Fig. 5) which is software that allows both the modeling of any object in three dimensions and the animation of those objects.

As it was said before, the chosen development engine is the Unreal Engine 4. To create the IVR application, we started with a template from Unreal and the models created in 3D Studio MAX were added, the necessary programming for the interaction with them was elaborated. When finished, he obtains the pieces in the environment (Fig. 6) and interacts with them, taking them, turning them in his hand and seeing them from different points of view.

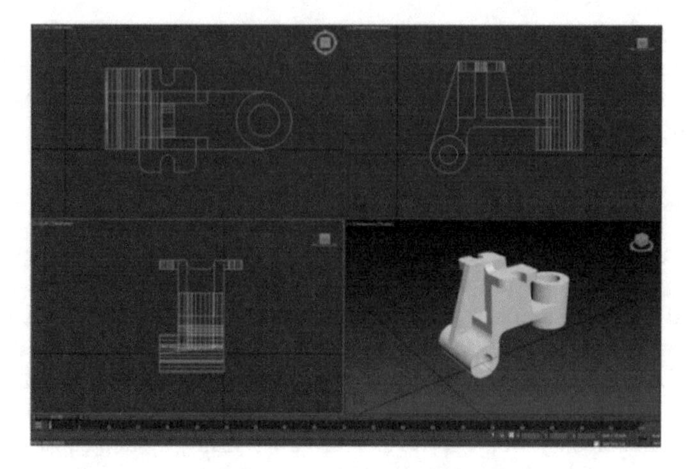

Fig. 5. One of the modeled parts.

Fig. 6. Point of view of the user, virtual hands and manipulation of a part.

3.4 Methodology of Use in the Classroom

A procedure for using the application in the subjects has been developed taking into account that not all students can use it. This methodology has the following phases:

- *Presentation of the program*: the first lesson comments on the importance of spatial skills in the professional performance of the engineer, as well as the academic performance of technical career students.
- *Initial evaluation*: the level of the spatial abilities of the students is evaluated by performing the spatial skills measurement tests at the beginning of the course: MRT [29] and PSVT:R [30].
- *Identification*: Students with lower level spatial abilities are identified from the tests.
- *Application*: students with poor test results use the virtual reality tool in three sessions of increasing complexity during the first two months of the course. This training has to be at the beginning of the course for the correct follow-up of the subject.

- *Evaluation*: by means of the administration at the end of the semester of the same tests used at the beginning, it is evaluated if there has been improvement in the spatial abilities.

The difference of results in the tests allows measuring the improvement in the spatial abilities of the students that use the application. In addition, this improvement can be compared with students who did not use it (control group).

4 Discussion

As indicated in the previous point, the use of this methodology should be done at the beginning of the subject, which could not be done this course since the process of creating the virtual reality tool in all its phases was laborious and occupied almost everything the time of the course.

The tool was only tested with some students to evaluate its operation and make the necessary corrections and that at the end of the course. It is expected to be able to apply this tool in the next academic course from the beginning of each subject and thus be able to evaluate the results.

5 Conclusions

We have been able to design and program a virtual reality computing tool making all the necessary studies of the state of the art and choosing the most suitable evaluation systems, hardware and software. Although the tests during the development allow to hope for an improvement of the spatial abilities of the students, the application in the subjects was not possible (it is expected to be able to do it in the next academic course).

Acknowledgments. This work has been developed as part of "Virtual-Ledgers-Tecnologías DLT/Blockchain y Cripto-IOT sobre organizaciones virtuales de agentes ligeros y su aplicación en la eficiencia en el transporte de última milla", ID SA267P18, project financed by Junta Castilla y León, Consejería de Educación, and FEDER funds.

References

1. Sorby, S.A.: Developing 3-D spatial visualization skills. Eng. Des. Graph. J. **63**(2), 21–32 (1999)
2. Rafi, A., Samsudin, K.A., Said, C.H.: Training in spatial visualization: the effects of training method and gender. Educ. Technol. Soc. **11**(3), 127–140 (2008)
3. Vergara, D., Lorenzo, M., Rubio, M.P.: Virtual environments in materials science and engineering: the students' opinion. In: Lim, H. (ed.) Handbook of Research on Recent Developments in Materials Science and Corrosion Engineering Education, 1st edn., pp. 148–165. IGI Global, Hershey (2015)
4. Adánez, G.P., Velasco, A.D.: Construção de um teste de visualização a partir da psicologia cognitiva. Avaliação Psicologica **1**(1), 39–47 (2002)

5. Uttal, D.H., Miller, D.I., Newcombe, N.S.: Exploring and enhancing spatial thinking: links to achievement in science, technology, engineering, and mathematics? Curr. Dir. Psychol. Sci. **22**(5), 367–373 (2013)
6. Wang, Ch.X, Zhao, Q., Sun, W., Wan, X., Cui, Q.: 3D scene of virtual reality system design and research. Key Eng. Mater. **522**, 761–768 (2012)
7. Vergara, D., Rubio, M.P., Lorenzo, M.: A virtual resource for enhancing the spatial comprehension of crystal lattices. Educ. Sci. **8**, 153 (2018)
8. Griol, D., Molina, J.: Measuring the differences between human-human and human-machine dialogs. ADCAIJ: Adv. Distrib. Comput. Artif. Intell. J. **4**, 2 (2015)
9. Palomino, C.G., Nunes, C.S., Silveira, R.A., González, S.R., Nakayama, M.K.: Adaptive agent-based environment model to enable the teacher to create an adaptive class. In: Advances in Intelligent Systems and Computing, vol. 617 (2017)
10. Chamoso, P., González-Briones, A., Rodríguez, S., Corchado, J.M.: Tendencies of technologies and platforms in smart cities: a state-of-the-art review. Wirel. Commun. Mob. Comput. **2018**, 17 (2018)
11. Gonzalez-Briones, A., Prieto, J., De La Prieta, F., Herrera-Viedma, E., Corchado, J.M.: Energy optimization using a case-based reasoning strategy. Sensors **18**(3), 865 (2018)
12. García, O., Chamoso, P., Prieto, J., Rodríguez, S., De La Prieta, F.: A serious game to reduce consumption in smart buildings. Commun. Comput. Inf. Sci. **722**, 481–493 (2017)
13. Vergara, D., Rubio, M.P., Lorenzo, M.: On the design of virtual reality learning environments in engineering. Multimodal Technol. Interact. **1**, 11 (2017)
14. Vergara, D., Rubio, M.P., Lorenzo, M.: Interactive virtual platform for simulating a concrete compression test. Key Eng. Mater. **572**, 582–585 (2014)
15. Vergara, D., Rubio, M.P.: Active methodologies through interdisciplinary teaching links: industrial radiography and technical drawing. J. Mater. Educ. **34**(5–6), 175–186 (2012)
16. Vergara, D., Rubio, M.P., Lorenzo, M.: New approach for the teaching of concrete compression tests in large groups of engineering students. J. Prof. Issues Eng. Educ. Pract. **143**(2), 05016009 (2017)
17. Sampaio, A.Z.: Virtual reality technology applied in teaching and research in civil engineering education. J. Inf. Tech. Appl. Educ. **1**, 152–163 (2012)
18. Chou, Ch., Hsu, H.-L., Yao, Y.-S.: Construction of a virtual reality learning environment for teaching structural analysis. Comput. Appl. Eng. Educ. **5**, 223–230 (1997)
19. Vergara, D., Rubio, M.P., Lorenzo, M.: A virtual environment for enhancing the understanding of ternary phase diagrams. J. Mater. Educ. **37**(3–4), 93–101 (2015)
20. Vergara, D., Rodríguez, M., Rubio, M.P., Ferrer, J., Núñez, F.J., Moralejo, L.: Formación de personal técnico en ensayos no destructivos por ultrasonidos mediante realidad virtual. Dyna **93**(2), 150–154 (2018)
21. Rubio, M.P., Vergara, D., Rodríguez, S., Extremera, J.: Virtual reality learning environments in materials engineering: rockwell hardness test. In: Di Mascio, T. et al. (eds.) Methodologies and Intelligent Systems for Technology Enhanced Learning (MIS4TEL 2018), AISC 804, pp. 106–111. Springer, Switzerland (2019)
22. Boletsis, C.: The new era of virtual reality locomotion: a systematic literature review of techniques and a proposed typology. Multimodal Technol. Interact. **1**, 24 (2017)
23. Bhattacharjee, D., Paul, A., Kim, J.H., Karthigaikumar, P.: An immersive learning model using evolutionary learning. Comput. Electr. Eng. **65**, 236–249 (2018)
24. De Freitas, S., Rebolledo-Mendez, G., Liarokapis, F., Magoulas, G., Poulovassilis, A.: Learning as immersive experiences: using the four-dimensional framework for designing and evaluating immersive learning experiences in a virtual world. Br. J. Educ. Technol. **41**, 69–85 (2010)

25. Lee, E.A.L., Wong, K.W.: Learning with desktop virtual reality: low spatial ability learners are more positively affected. Comput. Educ. **79**, 49–58 (2014)
26. Parong, J., Mayer, R.E.: Learning science in immersive virtual reality. J. Educ. Psychol. **110**(6), 785–797 (2018)
27. Conference for AR & VR innovation: http://www.xrdconf.com. Accessed 08 Feb 2019
28. Unreal Engine 4 Games: https://wiki.unrealengine.com/Category:Games. Accessed 08 Feb 2019
29. Vandenberg, S.G., Kuse, A.R.: Mental rotations, a group test of three-dimensional spatial visualization. Percept. Mot. Ski. **47**(2), 599–604 (1978)
30. Guay, R.B.: Purdue Spatial Visualization Tests. Purdue Research Foundation, West Lafayette (1977)

Learning and Development Is the Key. How Well Are Companies Doing to Facilitate Employees' Learning?

Leonardo Caporarello, Beatrice Manzoni(✉), and Beatrice Panariello

SDA Bocconi School of Management, Bocconi University,
Via Bocconi 8, 20136 Milan, Italy
{leonardo.caporarello,
beatrice.manzoni}@unibocconi.it,
beatrice.panariello@studbocconi.it

Abstract. Employees value learning and development as key factors for attraction, retention and engagement. Yet organizations are not always doing enough in order to support and facilitate employees' learning through the so-called organizational learning mechanisms. In this paper, we explore how employees perceive and evaluate their company's efforts in creating and implementing these mechanisms. We surveyed 247 employees and we discovered that while employees' satisfaction and enjoyment towards learning opportunities are high, learning and development initiatives are often not enough aligned to the individual needs as much as they are to the company's ones. Moreover, respondents suggest that learning sometimes remains exclusively on paper, either because the execution of learning activities takes too long or because, once they get back to work, employees do not get enough feedback afterwards. These findings offer to organizations a set of recommendations to provide more effective learning experiences.

Keywords: Organizational learning mechanisms · Learning and development · Employees · Organizations · HR

1 Introduction

Creating learning experiences that matter for both the employees and their organizations is widely recognized as fundamental, both by researchers [10, 13, 20] and practitioners [19]. Scholars are interested in exploring how organizations can enable its organizational members' learning in order to generate positive outcomes, such as satisfaction, improved performance, innovation, efficiency, and last but not least, competitive advantage.

Yet it is not always the case that organizations manage to provide meaningful learning experiences, especially if we listen to the voice of the employees and also to the one of the senior executives [19]. This often occurs because organizations tend to fail in aligning learning initiatives to business needs and in providing the proper organizational support to connect learning to employees' role responsibilities and career plans [19]. They also sometimes fail in creating a supportive learning-oriented culture [4].

R. Gennari et al. (Eds.): MIS4TEL 2019, AISC 1007, pp. 80–88, 2020.
https://doi.org/10.1007/978-3-030-23990-9_10

Existing literature grouped these issues into the concept of organizational learning mechanisms [2], intended as a set of organizational values, processes and systems that support and facilitate individual and organizational learning. Over the years, scholars have studied and categorized the different types of organizational learning mechanisms as well as their impact in terms of individual and organizational outcomes. Yet, we still know very little about how employees perceive and evaluate their company's efforts in creating and implementing these mechanisms.

In this article, we describe how people perceive and evaluate their company's efforts in terms of aligning learning to organizational goals and providing the contextual support to make it relevant for both individuals and the organization (organizational learning mechanisms).

2 Organizational Learning Mechanisms (OLMs)

Organizations can - and have the responsibility to - support and facilitate employees' learning. It is not new that in the absence of explicit intention and appropriate mechanisms, the learning potential may be lost. Organizations can enhance employees' learning by adopting a set of OLMs, which are those organizational processes and structures that can create or improve learning opportunities [2] and help organizational members to gather and apply knowledge-related resources effectively [16].

Given this, it is undoubtedly clear that OLMs play a fundamental role in organizations [3]. For example, they generate positive outcomes related to knowledge creation [6], continuous improvement [14], the fostering of creative environment [7], and organizational performance [11].

Existing literature has categorized OLMs into cultural and structural facets [2, 12, 16, 17]. Cultural facets, or cognitive mechanisms [7] enable the development of a learning culture. These include having a shared vision and values, norms, assumptions and beliefs, roles and behaviors. Structural facets, which include also procedural mechanisms [7], are related to people development processes, as well as to all those elements ensuring that learning activities are supported and realized within the workplace. For example, they include leadership and management, change management, communication, information and knowledge systems, performance management, and technology.

All in all, these studies posit OLMs as extremely relevant to sustain the organizational competitive advantage. At the same time, while research often explores 'why' setting OLMs in place is particularly important within an organization and 'what' precisely OLMs are, we still know little about how employees perceive their implementation within their employing organization.

3 Methods

This study is part of a broader research on how we learn and how we expect to learn within organizations in the future.

For the purpose of this study the sample consists of 247 employees. 52,6% of the respondents are male, 98% are Italian and over 95% are currently working in Italy. In terms of age the majority of participants reported to be between the ages of 26 and 45, which stands at 66.4% as age category. Average age is 38 years old (SD 10,57). In terms of functional role, 47,8% of the respondents work in HR - mainly in training and development (76,3%); the other half is well distributed among the other organizational functions. In terms of industry, 50,2% of the respondents works in the service industry, 19,4% in the manufacturing industry, 11,3% in the public government industry and 19% in others, not specified. Half of the sample works for big companies (more than 250 employees). The sample is nearly perfectly split between those who work for major companies (more than 250 employees) and those who reported to be employed at small and medium companies (less than 250 employees). In terms of seniority, the sample is nearly perfectly split between those who reported to have been working for more than 10 years now and those below this threshold. Average seniority stands at 14 years (SD 10,17). In terms of job level, 65,2% of the sample are managers (senior, middle or junior), while the rest of them are professionals with no specific managerial responsibility.

Applying a CAWI methodology, we measured the following scales from Armstrong and Foley [2], asking respondents to rate the extent to which they agreed to each and every statement (from 1 = Strongly disagree to 5 = Strongly agree, "N/A or I don't know"), unless differently explained: Mission linked learning; Facilitative learning environment; Learning identification satisfaction; Learning and development need (organizational support); Learning application; Learning satisfaction. For Learning enjoyment, we used a scale from Lin et al. (2008). For all these measures the values of Cronbach's alpha are >0,70.

4 Results

In this section we present the main findings of current study[1].

4.1 Employees Are Satisfied and Enjoyed the Learning Opportunities Their Companies Provide Them with

Enjoyment towards learning and self-development appears to be a widely spread attitude: 85% report to be absorbed and extremely focused while learning and 93% feel happy and satisfied during the learning experience, which is moreover meaningful and worthwhile to almost the entire sample (92%). Learning programs result also to be satisfying in meeting the learning needs of 75% of the respondents, even if 35% believe that the learning pace does not match the learning needs, sometimes even clashing with work demands (Fig. 1).

[1] Full tables with all the data are available upon request contacting the authors.

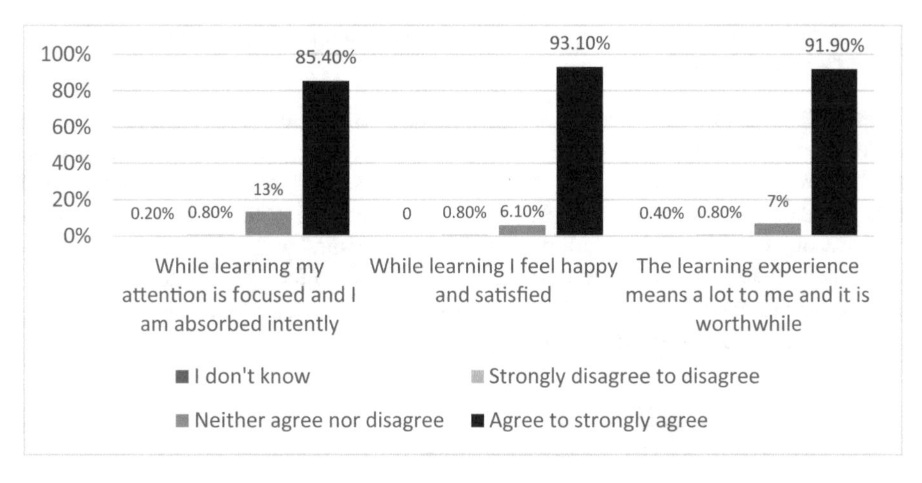

Fig. 1. Enjoyment towards learning.

4.2 Learning and Development Is Perceived as More Employer-Oriented Than Employee-Oriented

Employees seem to perceive that their companies are investing a lot on learning which is relevant for the organization. 81% of the overall respondents said that they believe that learning and development goals are highly connected to their organizations' mission and vision and, according to 75% of the sample, their companies value the development of the staff as fundamental to the overall organizational success. Moreover, 72% feel encouraged to improve themselves and develop their full potential.

Yet, 35% or more do not fully agree upon the fact that companies are always able to identify the correct resources to be used to meet individuals' training and development needs and to have regular processes for reviewing employees' training needs.

Organizations are perceived on the one hand as highly effective in developing learning opportunities that meet business needs, on the other hand as less effective in tailoring individual learning opportunities matching individual needs.

4.3 Employees Believe They Can Contribute Much More Than What Companies Let Them Do

86% of the total respondents feel confident that they own the necessary skills and knowledge to contribute to the development of their company, even if the percentage slightly drops when it comes to be fairly recognized as effective contributors to organizational performance (67%).

At the same time, employees believe their top management does not always understand the broader benefits and costs of development initiatives (36%) and, according to nearly 40% of the sample, its commitment to employees' continuous development is not clearly communicated throughout all the organizational levels.

4.4 Employees Know Which Skills They Need Despite What Their Organizations Do to Develop Them

When referring to their specific work unit, 80% of respondents declare to clearly understand what skills and knowledge they need to do their job in a proper way, while they believe the organization is not always equally aware of them. In fact, 36% think in fact that the skills of existing employees aren't developed in line with predetermined business objectives; and 41% report not to be fully satisfied with how their learning and development needs are currently being identified, offering chances for further engagement in decisional processes. Up to date, 60% take active part into staff training, learning and development decisions (Fig. 2).

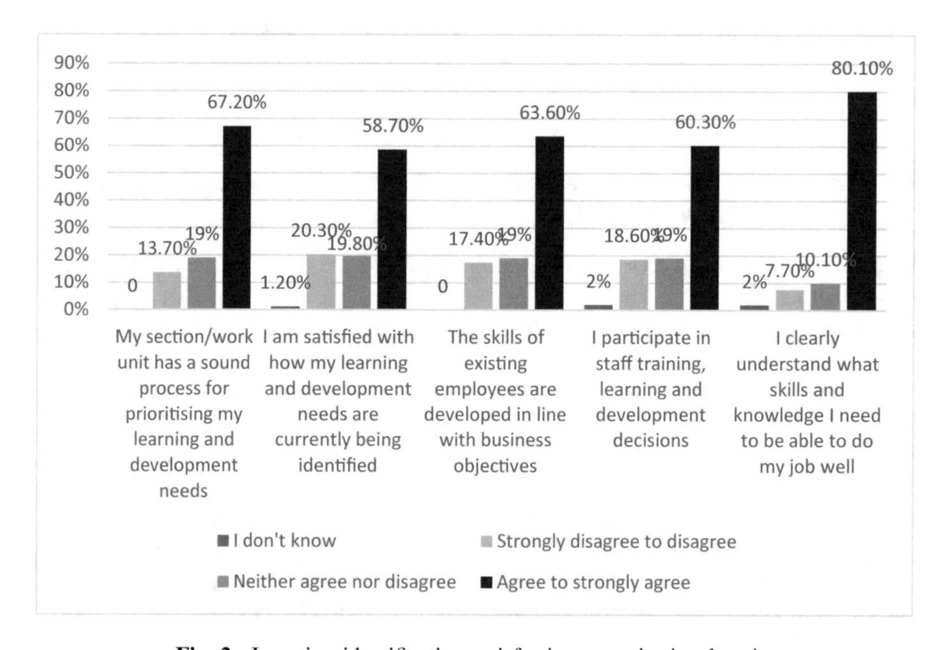

Fig. 2. Learning identification satisfaction – section/work unit.

4.5 Employees See Their Bosses as Effective Learning and Development "Partners"

The majority of respondents (68%) reports to be satisfied with the relationship with their bosses with regard to their learning and development plan. 71% recognize that their supervisors apply a constructive approach when discussing mutual learning and development needs, as well as show a clear commitment in meeting their requests, and that they perfectly agree on which their learning and development needs are (67%).

72% feel encouraged by their supervisor to undertake activities aiming at improving their knowledge and skills and 70% believe that learning and development opportunities are made available to all the members of the work unit. Meeting their supervisor at least once a year to discuss learning needs is a common practice for 68% respondents, even if around 38% don't have the chance to discuss how their job will change in the next future (Fig. 3).

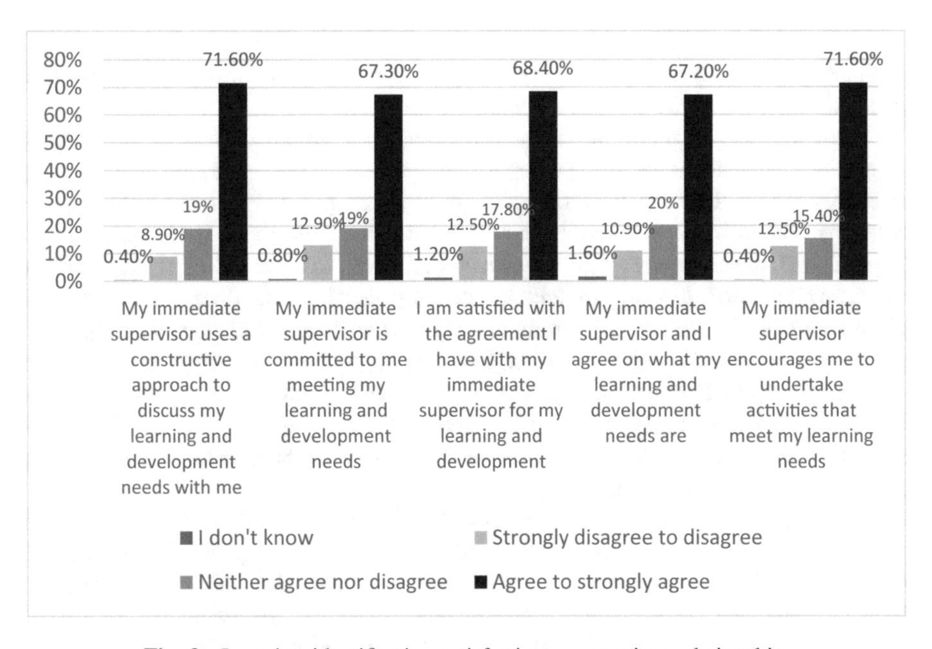

Fig. 3. Learning identification satisfaction – supervisor relationship.

4.6 Employees Perceive the Organization Is not Quick Enough in "Executing" Learning and Development Programs as Agreed

Employees report to have a clear understanding of how this kind of activities help them in increasing effectiveness on the job (75%) and to have access to available learning opportunities (68%). However, they complain about the fact that they cannot take part into learning experiences in the short term (44%). While the analysis of the training needs doesn't take too long, the learning experience itself might take longer to occur – at least according to what employees aim for.

4.7 There Is Room for Applying New Learning into Daily Work Life, yet Feedback Is not Enough

Overall, almost 80% of the sample feel encouraged to share what they have learnt with co-workers of the same unit and nearly 59% are usually asked to evaluate the eventual suitability of the development activities for the rest of the team as well. 67% say they have time to practice what they have learnt in the workplace. 70% usually discuss with their supervisor how to apply to the job what they have learnt and 65% feel that him or her helps them to put their learning into practice. The issue is that nearly 40% don't receive any feedback on how well they are using what they got from completed activities.

5 Conclusions

With this paper, we aim to provide both a research-oriented contribution and a practice-oriented one in the field of learning within organizations.

From a research point of view, we provide an empirical test of the OLMs' dimensions, investigating the perception employees have regarding the learning mechanisms that their organizations commonly put in place. The results of the questionnaire demonstrate that people enjoy and are well aware of the importance of learning and development at an organizational level. Yet, most of them claim an increasing attention towards individual learning needs, demanding a more tailored approach to HRD. Moreover, a strong and significant relationship between employees and their supervisors emerges. This finding also supports the literature that discusses the impact of job training on psychological variables such as motivation and commitment [5, 15], organizational trust [1], and willingness to go above and beyond to meet their job requirements [8], as well the importance of the training environment in enhancing the learning experience [18].

From a practice point of view, we presented HR professionals insights for a better design of the learning experiences they offered to their employees. First of all, HR should better align organizational instances with individual ones, at least ensuring that people could have better visibility and understanding of the overall HR processes. While employees are satisfied for what their line manager does to define their training needs, in terms of selecting initiatives and creating space for applying newly acquired skills, they are dissatisfied in particular with regard to what HR members do in terms of training needs' assessment. Employees (and their bosses) are well aware of prospective development areas, while HR staff doesn't. Secondly, learning sometimes remains on paper, either because execution of learning opportunities takes too long or because back to work people do not get enough feedback on their "new" way of doing things. This issue deals with the perennial issue of measuring the return on training investment. However, organizations should exploit the positive relationship employees declare to have with their bosses creating the culture for an ongoing informal feedback. Considering the structure of the study, a list of limitations to this research must be acknowledged. First of all, survey research carries with itself several problems associated with its use. Being surveys self-reported instruments, they may not always be completely valid or reliable. However, it can be reported that in this case a strong internal consistency of the instrument was confirmed by using Cronbach's alpha coefficient, providing reliability information about the ordinal data that have been collected. It is still important to state that this research is one of only a few attempting to investigate employees' point of view on Organizational Learning Mechanisms. Based on these results, future research can focus on investigating how learning and development activities can be designed to act as an effective mediator to the needs of both the individual and the organization. Further research can be directed towards development of appropriate methods to provide a comprehensive assessment of training results.

References

1. Anschutz, E.E.: TQM America: How America's Most Successful Companies Profit from Total Quality Management. McGuinn & McGuire, Bradenton (1995)
2. Armstrong, A., Foley, P.: Foundations for a learning organization: organization learning mechanisms. Learn. Organ. **10**(2), 74–82 (2003)
3. Beer, M.: Developing an effective organization: intervention method, empirical evidence, and theory. Res. Organ. Chang. Dev. **19**, 1–54 (2011)
4. Cannon, M.D., Edmondson, A.C.: Failing to learn and learning to fail (intelligently): how great organizations put failure to work to innovate and improve. Long Range Plan. **38**(3), 299–319 (2005)
5. Cheng, E.W., Ho, D.C.: A review of transfer of training studies in the past decade. Pers. Rev. **30**(1), 102–118 (2001)
6. Chou, S., Wang, S.: Quantifying 'ba': an investigation of the variables that are pertinent to knowledge creation. J. Inf. Sci. **29**, 167–180 (2003)
7. Cirella, S., Canterino, F., Guerci, M., Shani, A.B.: Organizational learning mechanisms and creative climate: insights from an Italian fashion design company. Creat. Innov. Manag. **25** (2), 211–222 (2016)
8. Rowden, R.W., Connie, C.T.: The impact of workplace learning on job satisfaction in small US commercial banks. J. Work. Learn. **17**(4), 215–230 (2005)
9. Ferrazzi, K.: Use your Staff Meeting for Peer to Peer Coaching. Harvard Business Review, February 2015
10. Fink, L.D.: Creating Significant Learning Experiences. Jossey-Bass, San Francisco (2013)
11. Fredberg, T., Norrgren, F., Shani, A.B.(.R.).: Developing and sustaining change capability via learning mechanisms: a longitudinal perspective on transformation. In: Woodman, R., Pasmore, W., Shani, A.B.(.R.). (eds.) Research in Organizational Change and Development, vol. 19, pp. 117–161. Emerald Publications, Bingley (2011)
12. Gephart, M., Marsick, V.: Learning organizations come alive. Train. Dev. **50**(12), 34–46 (1996)
13. Huysman, M.: Balancing biases: a critical review of the literature on organizational learning. In: Easterby-Smith, M., Burgoyne, J., Araujo, L. (eds.) Organizational Learning and the Learning Organization: Developments in Theory and Practice. Sage, London (1999)
14. Oliver, J.: Continuous improvement: role of organizational learning mechanisms. Int. J. Qual. Reliab. Manag. **26**, 546–563 (2009)
15. Pool, S., Pool, B.: A management development model: measuring organizational commitment and its impact on job satisfaction among executives in a learning organization. J. Manag. Dev. **26**(4), 353–369 (2007)
16. Popper, M., Lipshitz, R.: Organizational learning mechanisms: a cultural and structural approach to organizational learning. J. Appl. Behav. Sci. **34**, 161–179 (1998)
17. Popper, M., Lipshitz, R.: Organizational learning: mechanisms, culture, and feasibility. Manag. Learn. **31**, 181–196 (2000)

18. Santos, A., Stuart, M.: Employee perceptions and their influence on training effectiveness. Hum. Resour. Manag. J. **13**, 27–45 (2006)
19. Spar, B., Dye, C., Lefkowitz, R., Pate, D.: Workplace Learning Report, LinkedIn Learning (2018)
20. Yoon, S.: In search of meaningful online experiences. New Dir. Adult Contin. Educ. **100**, 19–30 (2003)

Investigating Gamification and Learning Analytics Tools for Promoting and Measuring Communities of Inquiry in Moodle Courses

Maria Tzelepi[1(✉)], Ioannis Petroulis[1], and Kyparisia Papanikolaou[2]

[1] National and Kapodistrian University of Athens, Athens, Greece
tzelepimaria@yahoo.com, johnyend@di.uoa.gr
[2] School of Pedagogical and Technological Education, Athens, Greece
kpapanikolaou@aspete.gr

Abstract. This article proposes the use of gamification techniques to motivate students and cultivate the sense of a Community of Inquiry (CoI) through asynchronous discussions. Prior research is being examined to identify essential factors in relation to Communities of Inquiry. These factors provide key insights for the development of a CoI through the asynchronous discussions and thus, they guide the selection of learning analytic tools and gamification mechanisms within the course. In this article it is proposed that measuring and exploring the aforementioned factors by the instructors through learning analytics for rewarding good social practice through gamification, promotes cognitive presence. This process is relevant to educators interested in developing critical thinking skills and may serve to support them redesign the course or teaching interventions.

Keywords: Communities of Inquiry · Gamification · Learning analytics · Moodle

1 Introduction

The Community of Inquiry (CoI) [1] is a widely accepted model used in online learning, for outlining learning experience aiming to the development of critical thinking, one of the 21st-century core skills.

Despite CoI's widespread application, its evaluation still remains challenging. The way CoI is evaluated in a learning context usually follows the two approaches (a) the qualitative content analysis for assessing the quality of the discussions taking place in the community as well as (b) the analysis of the CoI questionnaire [2] given to the learners at the end of the course. Specifically, both of them, the qualitative content analysis which is performed manually as well as the statistical analysis of the questionnaires, they are time-consuming and above all, they take place after the end of the course.

Efforts have been made to overcome these difficulties by exploiting technologies for (semi) automated content analysis so that the researcher does not endure this process [3]. However, these attempts have not yet achieved an acceptable accuracy so

R. Gennari et al. (Eds.): MIS4TEL 2019, AISC 1007, pp. 89–96, 2020.
https://doi.org/10.1007/978-3-030-23990-9_11

that the results could contribute to reliable conclusions. In addition, learning analytic (LA) techniques have been utilized during the learning process for researching potential correlation between LA data and CoI, but although the research field is particularly active [3], there is still a lot to be done in this direction.

There is a growing number of courses delivered using e-learning environments, especially in higher education, such as Moodle and Blackboard where online asynchronous discussions play an important role. Asynchronous discussion activities aim to actively engage learners in sharing information and perspectives by interacting with their peers [4].

Given that most activities of learners enrolled in online courses occur in the Learning Management System (LMS), utilizing the log data within the LMS could provide crucial insight into learners' behavior throughout the learning process [5]. If we can distinguish the learning patterns in the early stage of an online course, it will be conducive to encouraging or guiding learners through an appropriate instructional intervention [6]. In this direction, gamification has emerged as an effective teaching aid aiming to motivate learners to engage in the learning process [7].

In this paper, we focus on using learning analytics data in order to provide awards through gamification mechanisms. This way we aim at promoting best practices that learners adopt throughout online discussions and guide them to higher levels of cognitive presence.

2 Theoretical Background

2.1 Communities of Inquiry

Learning, according to the model of Community of Inquiry (CoI), occurs within the community through the interaction of three equally important elements. These elements are: **cognitive presence**, **social presence** and **teaching presence**. The participants of the community in which the deep and meaningful learning takes place are the educator that guides the learning process and helps learners not to deviate from its borders and of course, learners, who are the most significant part of the learning process [8]. Each of the abovementioned elements has indicators and a reference to them is made below.

Cognitive presence appears in any process that takes place in a CoI enabling learners to construct new knowledge and understand sufficiently every aspect of the discussion in which they participate while communicating with one another [9]. Cognitive presence is included in critical thinking, which is frequently presented as the ulterior goal of higher education and its process is described by sequence of the **triggering event**, the **exploration**, the **integration** and the **resolution** phase [10].

We can describe **social presence** as the ability of learners to project their personal characteristics into the communication, while presenting themselves to the rest of the class. Social presence is divided in three categories: **effective expression**, **open communication** and **group cohesion** [10].

As for the **teaching presence**, it deals with the designing and managing of the learning process, providing facilitation to learners in order to keep participating in an active way through all the activities of the learning process. It consists of **design**,

organization and leadership (**facilitation of discourse** and **direct instructions**), which are the categories of the teaching presence [10].

2.2 Gamification in Online Learning Systems

According to Kapp, gamification is the use of game like mechanisms, aesthetics and thinking into non-gaming environments or processes (such as education) to engage people, motivate them, promote learning and even, solve problems [11]. Gamification, as a method for improving people's motivation and engagement, is based on some gaming features, such as [7]: (a) the users' participation in a story or in general, pre-structured process, (b) challenges/tasks that users perform, (c) points that users collect, (d) levels that users conquer, (e) badges awarded to users and (f) ranking of users based on their achievements.

According to the literature, the use of gamification mechanisms improves the abilities to learn new skills by 40% [12]. Gamification affects learners' behaviour and attitude towards learning, improving their motivation and engagement. It finally provokes improvement of learners' knowledge level and skills, creating at the same time, conditions for an effective learning process [13].

In this context, as Huang and Soman [12] have mentioned, two categories of activities that demand different rewords can be defined (a) activities that require independent work and they should have as a reword individual awards (such as badges visible or invisible to the rest of learners) and (b) activities requiring interaction with fellow learners, where the social presence makes its appearance inside the learning process, and learners act as a part of a learning community, should be designed in a way that learners' results can be visible from all course participants (badges visible to the rest of learners) [13]. What makes the second category of activities so interesting and important for the learning process is, according Ehlers, the fact that the learning process is more effective if it has derived from collaboration and constructive discussion [14]. Online courses give the opportunity to learners and educators to collaborate and communicate with one another, exchanging ideas and creating a global view of the subject to be taught, thus, enhancing their social engagement [15].

Gamification plugins designed and implemented into LMSs, a well-known example of which is Moodle, constitute an organized process of providing gamification elements to course participants. Many of these plugins, such as the "**Stamp collection**" and "**Level up!**" Moodle plugins, can substantially contribute to an online learning community, reinforcing the interactions between course participants, improving their socialization and making the communication process and the collaboration between them more effective and enjoyable [16].

3 Combining Gamification with Learning Analytics in an Online Course

According to Garrison [10], the purpose of social presence in an educational context is to create the appropriate conditions for inquiry and quality interaction (reflective and threaded discussions) in order learners to collaboratively achieve worthwhile educational goals. Learners' interaction and participation are two noteworthy quantitative factors that have been researched in relation to CoI presences. Regarding interaction, there is evidence that promoting student – student interaction may offer valuable benefits to learning outcomes [17, 18] thus, it has been proposed to promote learners' replies to their peers' messages rather than just stating their own opinion.

Regarding participation, many studies have revealed relationship between this factor (number of messages posted, period of navigation, times of access, etc.) and CoI [19].

Teaching presence brings in contact cognitive and social presence constructing an effective learning process [9]. It seems that Shea et al. suggest that the course with a higher teaching presence provides a better support for meaningful interactions to occur [35], leading towards a higher quality discourse, and increased learners' social presence.

Moreover, to establish a high level of cognitive presence in asynchronous online discussions, teaching presence needs to provide learners with externally-facilitated regulation scaffolding participation and interaction. Some of the critical factors of teaching presence for promoting social and consequently cognitive presence are: the appropriate course structure, the instructional leadership role of the instructors [20, 21] guiding questions, an introductory discussion, role assignment, the frequent and continuous participation to the forum and contribution in each phase of cognitive presence, establishing rules that avoid delaying replies, appropriate selection of discussion topics [22], etc.

Although high participation and interaction are both necessary prerequisites to succeed a high cognitive presence, they are still not sufficient. Therefore, other factors, such as engagement in using different learning resources and tools, have also been referred to promote inquiry-based learning activities [23].

In this line of reasoning, LMSs offer instructors and course designers the opportunity to appropriately analyze the user footprints and use them to embed some triggering elements to raise learners' motivation and engagement throughout the learning process [24]. These elements could be badges, leaderboards, scores and the level up ability, which include the Gamification concept [24]. Users' footprints in online courses can be analyzed in detail using Learning Analytic (LA) tools, embedded on the environment of an LMS or standalone. Some LMSs, such as Moodle, have APIs that enable the interoperation with LA tools and the creation of plugins to embed LA functionalities [25].

Regarding our research, trying to base the course design on the theoretical base of CoI, we are going to evaluate the three presences of the CoI, during the learning process and on these three pillars (social presence, cognitive presence and teaching presence) we are going to focus while designing and gamifying our course. In order to do so, we are going to design badges to be attributed to learners and the community, in the context of the Gamified e-Learning Model, which main components are the user, the learning process, the goal, and the environment [15]. Our users are divided in two categories facilitators (educators) and learners, all participating in our learning environment built in Moodle platform and following the learning flow that we have organized, trying to reach our goal [15], which is the motivation of our learners to create a CoI and combining its presences, to acquire knowledge. This objective differentiates our research in process from other related researches in the field of gamification and LA [26, 27].

As far as individual learners are concerned, the badges are going to award learners for specific actions such as (a) for finishing a task assignment or a quiz, as a triggering event to increase their cognitive presence, (b) for interacting with their peers and (c) for participating to the course and specifically to the forum.

Participation badges will be also awarded to learners either by the educator or by their fellow learners, thus trying to increase both social and teaching presence. This badge category will be awarded to learners under circumstances that the educator or the fellow learner considers that are the appropriate, so as a sense of delight and surprise to be created for learners [28].

Having mentioned that our willing is to mostly gamify the part of the course that has to do with the social and teaching presences of our learners throughout the course, we have decided to also focus on learners' interaction. However, there are different ways of learners' interaction, such as cooperative, competitive and social [29]. In the environment of the course, the learners' *social interaction and collaboration* will be also promoted by awarding badges to them for actions inside the Moodle forum that will include posting questions, answering to fellow learners' questions and awarding badges to them when they have the appropriate argument to justify their move.

In order to achieve the abovementioned course design and trying to avoid negative impact of competitive mechanisms such as [28], we decided to use a Moodle plugin to help us throughout our awarding process. A Moodle plugin that award stamps (which are similar to badges) to learners, entitled "**Stamp collection**". "**Stamp collection**" that allows an educator to give stamps (i.e. picture with a comment) to learners so they collect these stamps. The activity can be used in many ways, such as motivative bonus marks, absence marks, certification records etc. [30]. The stamps awarded to a learner can be designed from the educator, so as to best reflect a message and meet the requirements of the learning design [31]. Another Moodle plugin that we are going to use is "**Level up!**", classifying our learners into levels with the aim of motivation them and activating them to learn.

Aiming to involve learners in inquiry-based activities we choose to offer both individual and community badges in order to foster learners' interaction, participation and motivation with the aim of promoting CoI. As far as Moodle courses are concerned, the factors that will be used in order to provide badges are addressed to the individual as a community member and also to all the community (see Table 1).

The LA tools that have been developed for the Moodle platform (LA Enriched Rubric, Engagement Analytics and, Engagement Analytics and Moodle log files) and a stand-alone tool (Gephi) have been chosen, as the information they provide expresses the CoI factors we intend to analyze. These tools are going to be used with the aim of exploring the factors mentioned in the Table 1, during the course, so as the educator to be able to intervene on time redesigning the course and providing the appropriate rewards to the members of the community. To gain insight into the applicability of social network analysis as a useful tool for understanding the dynamics of online learning through the CoI model we will use density and intensity metrics.

According to Shea et al. [35] network density measures derived from social network analysis can be a useful addition for understanding the development of social presence in online environments. The advantage of this method is that instead of

Table 1. CoI factors captured for award based on data provided by the LA tools

LA tools adopted	Data provided by the tool	CoI factors for award
LA enriched rubric moodle plugin [31]	The number of classmates a learner has interacted with, in the selected course module	Number of peers that the learner has interacted with awarded by *Social Badges* for the learner
Engagement analytics moodle plugin [32]	Learners' login activity, assessment submission activity, forum viewing and posting activity	The time and the frequency that the learner spends in the course is awarded by *Individual Badges* for the learner
Gephi [33]	Density	The density of learners' interactions is awarded by *Community Density Badges* for the community
	Intensity	The average time spent and the average posts made per week is awarded by *Community Interaction Badges* for the community
Moodle log files [34]	Forum log data	The continuous participation (e.g. 2 posts per day) is awarded by *Participation Badges* for the learner

spending many hours of analyzing the content of the transcript, network density measures could be automatically calculated with log files commonly found in learning management systems such as Moodle.

4 Future Plans

Aiming to contribute to the design of Moodle LMS courses based on CoI model, in this paper we propose a design intervention including technological as well as pedagogical aspects. Our aim is to empower learners through inquiry based activities by using learning analytics and gamification mechanisms.

Currently, the online-course developed based on the specific design rational is on progress at the spring semester. Our plans for future research focus on data collection that will allow us to evaluate the adequacy and efficiency of the gamification interventions as well as of the factors selected to be awarded. Moreover, we aim to evaluate how and to what extent the social and cognitive presences have been developed as a result of the gamified teaching interventions, according to evidence from previous studies applied in online learning environments [15].

Acknowledgement. This work has been (co-)financed by the Greek School of Pedagogical and Technological Education through the operational program "Research strengthening in ASPETE": "Enhancing Communities of Inquiry through Learning Analytics and Gamification".

References

1. Garrison, D., Anderson, T., Archer, W.: Critical inquiry in a text-based environment: computer conferencing in higher education. Internet High. Educ. **2**(2–3), 87–105 (1999)
2. Arbaugh, J.B., Cleveland-Innes, M., Diaz, S.R., Garrison, D.R., Ice, P., Richardson, J.C., Swan, K.P.: Developing a community of inquiry instrument: testing a measure of the community of inquiry framework using a multi-institutional sample. Internet High. Educ. **11** (3), 133–136 (2008)
3. Kovanović, V., Joksimović, S., Waters, Z., Gašević, D., Kitto, K., Hatala, M., Siemens, G.: Towards automated content analysis of discussion transcripts: a cognitive presence case. In: Proceedings of the Sixth International Conference on Learning Analytics and Knowledge, pp. 15–24. ACM (2016)
4. Erlin, B., Yusof, N., Rahman, A.A.: Analyzing online asynchronous discussion using content and social network analysis. In: Proceedings of the Ninth International Conference on Intelligent Systems Design and Applications (2009)
5. Il-Hyun, J., Dongho, K., Meehyun, Y.: Analyzing the log patterns of adult learners in LMS using learning analytics. In: Proceedings of the 4th International Conference on Learning Analytics and Knowledge. ACM (2014)
6. Brown, M.: Learning analytics: the coming third wave. EDUCAUSE Learning Initiative Brief 1.4, pp. 1–4 (2011)
7. Kiryakova, G., Angelova, N., Yordanova, L.: Gamification in education. In: Proceedings of 9th International Balkan Education and Science Conference (2014)
8. Rourke, L., Garrison, D.R., Anderson, T., Archer, W.: Assessing social presence in asynchronous text-based computer conferencing. J. Distance Educ. **14**, 50–71 (2001)
9. Garrison, D.R., Anderson, T., Archer, W.: Critical inquiry in text based environment: computer conferencing in higher education. Internet High. Educ. **2**(2–3), 87–105 (2000)
10. Garrison, D.R.: Online community of inquiry review: social, cognitive, and teaching presence issues. J. Asynchronous Learn. Netw. **11**(1), 61–72 (2007)
11. Kapp, K.M.: The Gamification of Learning and Instruction: Game-Based Methods and Strategies for Training and Education. Wiley, Hoboken (2012)
12. Huang, W.H.-Y., Soman, D.: A practitioner's guide to gamification of education. Rotman School of Management, University of Toronto, Toronto (2013)
13. Ehlers, U.D.: Web 2.0 e-learning 2.0 quality 2.0? Quality for new learning cultures. Qual. Assur. Educ. **17**, 296–314 (2009)
14. Waters, J., Gasson, S.: Distributed knowledge construction in an online community of inquiry. In: Proceedings of the 2007 40th Annual Hawaii International Conference on System Sciences (HISCC 2007), p. 200c (2007)
15. Utomo, A.Y., Amriani, A., Aji, A.F., Wahidah, F.R.N., Junus, K.M.: Gamified e-learning model based on community of inquiry. In: Proceedings of the 2014 International Conference on Advanced Computer Science and Information Systems (ICACSIS), Jakarta, Indonesia, pp. 474–480. IEEE (2014)
16. Bernard, R.M., Abrami, P.C., Borokhovski, E., Wade, C.A., Tamim, R.M., Surkes, M.A.: A meta-analysis of three types of interaction treatments in distance education. Rev. Educ. Res. **79**(3), 1243–1289 (2009). https://doi.org/10.3102/0034654309333844
17. Schrire, S.: Knowledge building in asynchronous discussion groups: going beyond quantitative analysis. Comput. Educ. **46**(1), 49–70 (2006)
18. Naranjo, M., Onrubia, J., Segués, M.T.: Participation and cognitive quality profiles in an online discussion forum. Br. J. Educ. Technol. **43**(2), 282–294 (2012)

19. Vrasidas, C., McIsaac, M.S.: Factors influencing interaction in an online course. Am. J. Distance Educ. **13**(3), 22–36 (1999)
20. Garrison, D.R., Cleveland-Innes, M.: Facilitating cognitive presence in online learning: interaction is not enough. Am. J. Distance Educ. **19**, 133–148 (2005)
21. Cheung, W.S., Hew, K.F., Ng, S.L.: Toward an understanding of why students contribute in asynchronous online discussions. J. Educ. Comput. Res. **38**(1), 29–50 (2008)
22. Akyol, Z., Garrison, D.R.: Understanding cognitive presence in an online and blended community of inquiry: assessing outcomes and processes for deep approaches to learning. Br. J. Educ. Technol. **42**(2), 233–250 (2011)
23. Akyol, Z., Garrison, D.R.: Assessing metacognition in an online community of inquiry. Internet High. Educ. **14**(3), 183–190 (2011)
24. Muntean, C.I.: Raising engagement in e-learning through gamification. In: Proceedings of the 9th International Conference on Virtual Learning (2011)
25. Katsigiannakis, V., Karagiannidis, C.: Research on e-Learning and ICT in Education. Springer, Cham (2017)
26. Klemke, B., Eradze, M., Antonaci, A.: The flipped MOOC: using gamification and learning analytics in MOOC design—a conceptual approach. Educ. Sci. **8**, 25 (2018)
27. Cassano, F., Piccinno, A., Roselli, T., Rossano, V.: Gamification and learning analytics to improve engagement in university courses. In: Di Mascio, T., et al. (eds.) 8th International Conference on Methodologies and Intelligent Systems for Technology Enhanced Learning. MIS4TEL 2018. Advances in Intelligent Systems and Computing, vol. 804. Springer, Cham (2019)
28. Lee, J.J., Hoadley, C.: Leveraging identity to make learning fun: possible selves and experiential learning in massively multilayer online games (MMOGs). J. Online Educ. **3**(6), 5 (2007)
29. Motivating students through Badges with the Stamp Collection module (Moodle 2.2.x). https://www.moodlenews.com/2012/motivating-students-through-badges-with-the-stamp-collection-module-moodle-2–2-x/
30. Activities: Stamp collection. https://moodle.org/plugins/mod_stampcoll. Accessed 06 Feb 2019
31. Grading Methods: Learning analytics enriched rubric. https://moodle.org/plugins/gradingform_erubric. Accessed 11 Feb 2019
32. Engagement analytics. https://moodle.org/plugins/report_engagement. Accessed 11 Feb 2019
33. Features. https://gephi.org/features/. Accessed 11 Feb 2019
34. Parise, P.: A preliminary look at online learner behavior: what can the moodle logs tell us? Bull. Kanagawa Prefect. Inst. Lang. Cult. Stud. **7**, 55–76 (2017)
35. Shea, P., Hayes, S., Vickers, J., Gozza-Cohen, M., Uzuner, S., Mehta, R., Rangan, P.: A re-examination of the community of inquiry framework: social network and content analysis. Internet High. Educ. **13**(1–2), 10–21 (2010)

Cognitive Emotions Recognition in e-Learning: Exploring the Role of Age Differences and Personality Traits

Berardina De Carolis[1(✉)], Francesca D'Errico[2], Marinella Paciello[3], and Giuseppe Palestra[4]

[1] Department of Computer Science, University of Bari, Bari, Italy
berardina.decarolis@uniba.it
[2] Fil.Co.Spe Department, RomaTre University, Rome, Italy
[3] International Telematic University UNINETTUNO, Rome, Italy
[4] Hero Srl, Milan, Italy

Abstract. It is well known that emotions have a great impact on the learning process and this becomes especially important when moving to on-line education. Then, endowing e-learning systems with the capability of assessing the emotional state of learners, can be used to provide feedback about their difficulties and problems. In this paper, we present an empirical study performed with a group of first-year students aiming at getting information on users' affective state during the learning process considering their personality traits. At this aim, we developed a tool for cognitive emotion recognition from facial expressions. Results show how detected emotions can be considered as an indicator of the e-learning process quality. Furthermore, another result is that cognitive emotions, experienced during e-learning process, can be strongly differentiated according to the learning activities, students age and personality.

Keywords: Emotions · E-learning · Facial expression recognition

1 Introduction

Affect plays a critical role in learning performances as it influences cognitive processes [1, 2]. A range of different emotions occur during the learning process, from positive, in case of successful achievement, to negative ones, as a consequence of failure and lack of understanding, to emotions related to interest, curiosity or perplexity in front of a new topic. The recognition of user's affective state may play an important role in improving the effectiveness of e-learning by providing a way to adapt the learning path and teaching style according to the recognized emotions that can be used as a feedback to the e-learning platform, thus enhancing the effectiveness of the e-learning environment [10, 11]. There are several theories that can be used to classify emotions, in the present paper we refer to a socio-cognitive approach [7] and appraisal theories [8], which define emotions as adaptive devices that monitor the state of achievement or thwarting of individuals' goals [9]. Beyond the emotional valence, the present work will explore the role played by a particular type of emotions, the cognitive ones. In this

R. Gennari et al. (Eds.): MIS4TEL 2019, AISC 1007, pp. 97–104, 2020.
https://doi.org/10.1007/978-3-030-23990-9_12

perspective, 'cognitive emotions', elicited to acquire or develop new skills/knowledges [5, 6], can play a crucial role because they indicate the state or the "flow" in the process. The state of flow is possible when students feel able to face also challenging tasks [12]. At this purpose, cognitive emotions are strictly monitoring the incoming content, can be considered 'strictly connected with the demands of consistency, order, clarity and relevance' [13]. Furthermore they can be considered as emotional filters through which we view the world, interpret its objects and evaluate its critical features [14] also by considering the way the content is presented, for the formal features of the content. This relationship between cognitive emotions and learning style is even more relevant in a modern learning context such as distance e-learning that uses different formats (video, chat, forum) to convey the content to teach. O'Regan [4], for instance, explored the experience of students learning online. The study identifies both positive and negative emotions experienced by students and their effect on the learning process.

Besides emotions, several studies have attested the importance to personality traits in predicting academic adjustment [26] and their relation to emotional states in academic situations. Researchers have found each of the Big Five personality traits (Openness, Conscientiousness, Extraversion, Agreeableness, Neuroticism) to be reliable predictors of academic performance [15]. However, taking this whole body of research into account, empirical literature reviews [24] and meta-analyses [15] identify Conscientiousness as the trait with the strongest and most consistent association with academic success. Some studies conducted on undergraduate college students find other traits (Openness and Agreeableness) to be significantly correlated with academic success [15].

To this aim, we performed an empirical study, at a Distance University, with a group of first-year students with the main goal of getting information on users' affective state during the learning process considering their personality traits. To analyze and recognize cognitive emotions we developed a tool for the recognition of emotions from facial expressions that typically arise during the learning process. Many systems proposed to use facial expression analysis for continuously and unobtrusively monitoring learners' behavior during e-learning and interpreting this into emotional states [16, 17]. However, they focused mainly on primary emotions recognition [3] that, in this application domain, are not sufficient since they do not allow for a deep understanding of user's mental state [4]. In this context emotions do not vary quickly and have a lower intensity than in other domains. For this reason, we decided to investigate whether cognitive emotions could be detected in this context and used to provide feedback to the e-learning system. Our tool called FEAtuREs (Facial Expressions Analysis for Recognition of Emotions) recognizes the following cognitive emotions: enthusiasm, interest, surprise, curiosity, concentration, attention, disappointment, boredom, perplexity, discomfort, and frustration.

The tool has been tested in a real context and used as a basis for a study aiming at: (i) detecting and recognizing cognitive emotions from facial expressions to ascertain cognitive emotions across two different, commonplace e-learning activities/situations: viewing prerecorded video lectures and participating in an online chat with a teacher/tutor, and (ii) exploring cognitive emotions in relation to personality traits in technology-mediated learning.

The paper, after a brief section describing the FEAtuREs tool and its accuracy in recognizing cognitive emotions, shows how it has been used in an experimental study by providing a discussion concerning the relation between cognitive emotions and e-learning settings and taking into account individual differences in terms of age and personality. Conclusions and future work directions are discussed in the last section.

2 The Cognitive Emotions Recognition System

As far as recognition of cognitive emotions is concerned, advances in the field of Affective Computing [2] have opened the possibility of recognizing emotions from non-verbal communication channels like facial expressions, body, voice and so on. In particular, the most informative channel for computer-based emotion recognition is the face [16]. The majority of these approaches interpret the emotional state of users during their interactions with an e-learning environment according to the six basic emotions. In distance education, it is feasible to assume that the recognition of emotions can happen through a webcam that is usually present in every computer and, when people attend at an online course, they are in front of the computer watching the provided material. Then, to measure cognitive emotions we developed an automatic detection system able to recognize facial expressions both from video and in real time during e-learning sessions. A robust Facial Expressions Recognition (FER) system should deal with the intrinsic variation of the same expression among different subjects in order to keep good performance with the unseen ones.

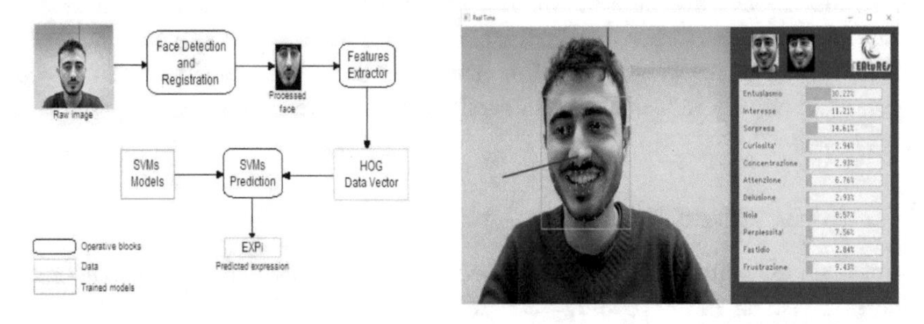

Fig. 1. (a) Pipeline of the proposed FER system. (b) Cognitive emotion recognition interface.

The tool (Fig. 1b), called FEAtuREs (Facial Expressions Analysis for Recognition of Emotions), follows a commonly used pipeline in FER systems [18]. It performs facial expression recognition on a single image and uses a global approach, considering as a region of interest the whole face. The set of descriptors used to recognize human emotion traits is based on the Histogram of Oriented Gradients (HOG) [19]. The pipeline (Fig. 1a) takes as input a single facial image, from the video stream, and the interface performs a preliminary face detection [20], it applies the HOG descriptors for features extraction step and finally classifies the facial expression by a multi class Support Vector Machines (SVMs) using a Radial Basis Function (RBF) as kernel with penalty parameter C = 1000 and $\gamma = 0.05$. In order to train and test the performance of the proposed approach we collected images coming from three different datasets: (i) EU-Emotion Stimulus Set [21], of the University of Cambridge; (ii) The Cambridge Mindreading Face-Voice Battery [22]; and The Cambridge Mindreading Face-Voice Battery for Children [23]. In particular, we selected eleven cognitive emotions that were mentioned in literature as relevant to the e-learning process: enthusiasm, interest, surprise, curiosity, concentration, attention, disappointment, boredom, perplexity, discomfort, frustration. The output of this selection is a set of 4184 images whose distribution is the following: enthusiasm (498), interest (340), surprise (295), curiosity (453), concentration (495), attention (374), disappointment (370), boredom (270), perplexity (369), discomfort (461), frustration (259).

The performance of the proposed approach have been analyzed by means of the k-fold cross validation with k = 10 and the average accuracy is of 92%. More in details we obtained the following results: TP Rate = 0.918, FP Rate = 0.08, Precision = 0.919, Recall = 0.918, F-Measure = 0.918. These are very encouraging results considering the challenging benchmark used for testing.

3 A Preliminary Study

The approach has been tested in the context of an empirical study performed at a Distance University of with a group of students of the Psychology Faculty. The study goal was twofold. On one side we wanted to test the reliability of trained software in detecting and recognizing cognitive emotions from facial expressions and, then, to analyze the emotional profile in two learning activities: video lectures and chat with teacher and relate this to personality traits. In particular, analyses will be carried out by considering the phase of students' life (emerging adult = under 30 yrs; adult = over 31 yrs) since mostly in a distance university students are usually heterogeneous with regard to age.

3.1 Participants and Procedure

Ten on-line university students (all females) took part in this study. They were enrolled in the first year of the Psychology course, aged between 20 and 64 (emerging adult students = 5, with age mean = 24.8 SD = 3.34; adult students = 5, with age mean = 52.4 SD = 11.5). The participants were previously invited to collaborate through dedicated mail by the researchers, students were then informed about the general purposes of the study, accepting to sign up an informed consent, along with the of study instructions. Participants fulfilled on-line questionnaires with their socio-demographic data and variable study and followed educational activities, taking care of one's own face in the foreground, through webcams during learning activities. At the end of their three videos collection they upload the material on a shared drive with invented acronyms in order to guarantee their anonymity. In total 20 videos have been collected which contained more than 28 h of recordings, ready to be automatically analyzed.

3.2 Material

To measure the personality traits of the students was used a short 298 version of Big Five Questionnaire [25] containing 60 items that assess big-five domain scales on a Likert scale 5 points (1 = very false for me; 5 = very true for me). Specifically, the scale assesses: Energy/Extraversion, referring characteristics such as activity, enthusiasm, assertiveness and self-confidence; Agreeableness referring characteristics such as concern and sensitivity towards others and their needs, and collaborative attitudes; Conscientiousness, referring assesses dependability, orderliness, precision, and the fulfilling of commitments. Emotional Instability (vs. neuroticism), referring the tendency to feel anxiety, depression, discontent, and anger. Finally, Intellect/Openness, referring tendency to explore new challenges and knowledge and to be creative. To measure cognitive emotions we used FEAtuREs.

3.3 Results: Emotional Profile Across e-Learning Settings

As shown in Fig. 2a, among the cognitive emotions expressed during the video-lectures, attention is the most frequent one, followed by boredom and frustration. This result is quite predictable, since in video-lectures the focalization phase takes a central role. But it is noticeable the fact that in this condition the young students are more attentive than adult ones. Adults during the video-lectures expressed more frustration and boredom.

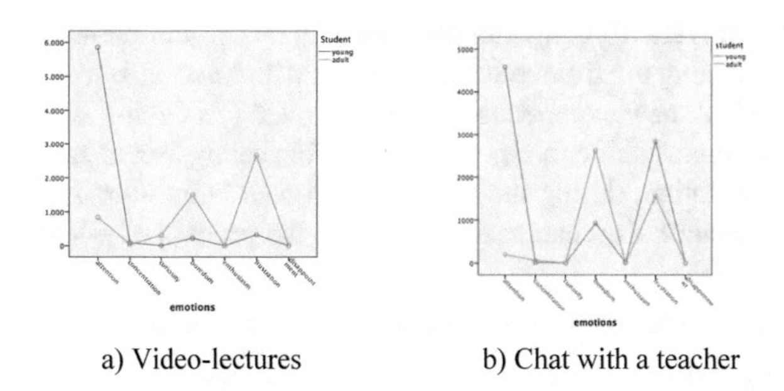

a) Video-lectures b) Chat with a teacher

Fig. 2. Emotional profiles in learning_condition * Age.

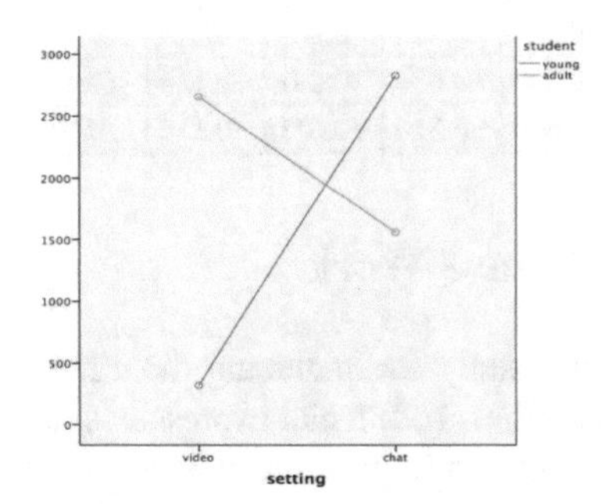

Fig. 3. Frustration * e-learning environments.

Also in the chat with teacher (see Fig. 2b) the expressed emotional profiles include mostly attention, frustration and boredom, but in this case when we check age differences we can see that young students express not only attention but more frustration than adult ones (Fig. 3). While young students express more frustration in chat with tutor, adult in the opposite condition (Fig. 3).

From the correlations reported below emerged how for young learners energy and openness to experience are negatively associated with perplexity during the video lectures (Table 1a), the higher energy and openness to experience trait the lower is the perplexity. For adults to the energy traits (Table 1b), the emotional stability is negatively correlated with boredom (absence of new stimuli) and frustration (critical stimuli).

Table 1. (a) Young learners vs. (b) Adult learners during video lectures.

BFQ	(a) Young learner					(b) Adult learner				
	IN	CU	BO	PE	FR	IN	CU	BO	PE	FR
Energy	0.514	−0.642	0.083	**−0.934***	0.176	−0.646	0.504	**0.862***	−0.040	**−0.859***
Agreeab.	0.470	−0.689	0.009	−0.741	0.104	−0.38	0.000	0.601	0.157	0.606
Consciu.	0.386	0.278	0.362	−0.174	0.333	−0.522	−0.179	−0.490	−0.545	−0.489
EmoSt	0.625	0.132	0.016	−0.298	0.021	−0.560	−0.198	**0.849***	−0.713	**−0.849***
Openess	0.711	−0.595	0.279	**−0.934***	0.364	−0.766	−0.293	0.114	−0.427	−0.119

The more extroverted young students are (Table 2a), the more they feel positive emotions in interacting environment like the chat with tutor, also in this case they live the chat as a positive new experience. In adult learners (Table 2b), we found again that energy and emotional stability are positive traits for feeling less negative emotions like boredom and frustration, during the interaction in chat with tutor. Adult learners with higher levels of neuroticism can easily exit from the process of e-learning flow.

Table 2. (a) Young learners vs. (b) Adult Learners during the chat with the tutor.

BFQ	(a) Young learner					(b) Adult learner				
	IN	CU	BO	PE	FR	IN	CU	BO	PE	FR
Energy	−0.684	−0.642	0.353	−0.834	−0.655	−0.430	−0.078	**−0.849**	0.654	**−0.846**
Agreeab.	−0.071	**0.811**	−0.755	−0.728	−0.731	−0.193	0.299	0.604	0.108	0.613
Consciu.	**−0.895***	−0.453	0.265	0.012	0.305	−0.379	**0.960****	−0.504	−0.230	−0.499
EmoSt	−0.444	0.767	0.023	−0.088	0.081	−0.136	0.593	**−0.859**	0.148	**−0.850**
Openess	−0.719	0.408	−0.660	**−0.810**	−0.611	0.116	0.276	0.117	0.621	0.140

4 Conclusions and Future Work

In this paper we described a study for analyzing the emotional profiles based on the recognition of cognitive emotions from facial expressions. Results are encouraging and seem to indicate that emotions can be used as indicator of the quality of the student's learning process.

From our results emerged how peculiar personality traits can predict the emotional experience, and in particular cognitive emotions, and that it can differ in young and adult students. From our results emerged how in young students, the energy and openness to experience can be a positive and protective factor for perplexity, mainly during the video lectures and chat with tutor. Young learners when are higher in the two agency dimensions (energy and openness) lived more positively the e-learning experience, as a new, and in some cases as 'futuristic', experience to live. Other personality traits predict the cognitive emotions of adult learners: in general, we found that energy is a key factor to avoiding a state of boredom and frustration across e-learning environments. In adults another important personality trait is the emotional stability, in the sense that neuroticism can be associated more easily with negative emotions that are provoked by an absence of new stimulus, as in the case of boredom, but also in presence of problematic notions, like in the case of frustration. Adults with a lack of emotional stability are highly correlated with negative emotions in all the three e-learning environments taken into consideration. Other trait that can be considered significant in adults is the Conscientiousness when they interact by chatting with peers, in this case the higher conscientious learners the lower negative emotions like frustration, perplexity and boredom are felt. We are aware that the experiment was conducted with a small number of subjects and, moreover they were all female, therefore in our future work will set up a new study involving a larger sample of students balanced for gender.

References

1. Damasio, A.R.: Descartes Error: Emotion, Reason and the Human Brain. G.P. Putnam Sons, New York (1994)
2. Picard, R.W.: Affective Computing. MIT Press, Cambridge (1997)
3. Castelfranchi, C.: Affective appraisal versus cognitive evaluation in social emotions and interactions. In: Affective Interactions, pp. 76–106 (2000)
4. O'Regan, K.: Emotion and e-learning. J. Asynchronous Learn. Netw. **7**(3), 78–92 (2003)
5. Castelfranchi, C., Miceli, M.: The cognitive-motivational compound of emotional experience. Emot. Rev. **1**(3), 223–231 (2009)
6. Scherer, K.R.: Psychological models of emotion. Neuropsychol. Emot. **137**(3), 137–162 (2000)
7. D'Errico, F., Poggi, I.: Social emotions. A challenge for sentiment analysis and user models. In: Tkalcic, M., De Carolis, B. (eds.) Emotions and Personality in Personalized Systems, pp. 13–34. Springer, Berlin (2016)
8. Feidakis, M., Daradoumis, T., Caballé, S., Conesa, J.: Embedding emotion awareness into e-learning environments. Int. J. Emer. Technol. Learn. **9**(7), 39–46 (2014)
9. D'Errico, F., Paciello, M., Cerniglia, L.: When emotions enhance students' engagement in e-learning processes. J. E-Learn. Knowl. Soc. **12**(4), 9–23 (2016)
10. Bassi, M., Steca, P., Delle Fave, A., Caprara, G.V.: Academic self-efficacy beliefs and quality of experience in learning. J. Youth Adolesc. **36**(3), 301–312 (2007)
11. Peters, R.S.: Moral Development and Moral Education. Routledge, New York (2015)
12. Israel, S.: In Praise of the Cognitive Emotions, p. 174. Routledge, New York (1991)

13. Sebe, N.: Multimodal interfaces: challenges and perspectives. J. Ambient Intell. Smart Environ. **1**, 23–30 (2009)
14. Khalfallah, J., Slama, J.B.H.: Facial expression recognition for intelligent tutoring systems in remote laboratories platform. Procedia Comput. Sci. **73**, 274–281 (2015)
15. Poropat, A.E.: A meta-analysis of the five-factor model of personality and academic performance. Psychol. Bull. **135**(2), 322 (2009)
16. De Carolis, B., de Gemmis, M., Lops, P., Palestra, G.: Recognizing users feedback from non-verbal communicative acts in conversational recommender systems. Pattern Recogn. Lett. **99**, 87–95 (2017)
17. Dalal, N., Triggs, B.: Histograms of oriented gradients for human detection. In: IEEE Computer Society Conference on Computer Vision and Pattern Recognition, CVPR 2005, vol. 1, pp. 886–893. IEEE, June 2005
18. Viola, P., Jones, M.: Robust real-time object detection. Int. J. Comput. Vis **57**(2), 137–154 (2004)
19. O'Reilly, H., Pigat, D., Fridenson, S., Berggren, S., Tal, S., Golan, O., Bölte, S., Baron-Cohen, S., Lundqvist, D.: The EU-emotion stimulus set: a validation study. Behav. Res. Methods **48**(2), 567–576 (2016)
20. Golan, O., Baron-Cohen, S., Hill, J.: The Cambridge mindreading (CAM) face-voice battery: testing complex emotion recognition in adults with and without Asperger syndrome. J. Autism Dev. Disord. **36**(2), 169–183 (2006)
21. Golan, O., Sinai-Gavrilov, Y., Baron-Cohen, S.: The Cambridge mindreading face-voice battery for children (CAM-C): complex emotion recognition in children with and without autism spectrum conditions. Mol. Autism **6**(1), 22 (2015)
22. Di Mele, L., D'Errico, F., Cerniglia, L., Cersosimo, M., Paciello, M.: Convinzioni di efficacia personale nella regolazione dell'apprendimento universitario mediato dalle tecnologie. Qwerty-Open Interdisc. J. Technol. Cult. Educ. **10**(2), 63–77 (2015)
23. Zeng, Z., Pantic, M., Roisman, G.I., Huang, T.S.: A survey of affect recognition methods: audio, visual, and spontaneous expressions. IEEE Trans. Pattern Anal. Mach. Intell. **31**(1), 39–58 (2009)
24. D'Errico, F., Paciello, M., De Carolis, B., Vattani, A., Palestra, G., Anzivino, G.: Cognitive emotions in e-learning processes and their potential relationship with students' academic adjustment. Int. J. Emot. Educ. **10**(1), 89–111 (2018)
25. Caprara, G.V., Barbaranelli, C., Borgogni, L.: BFQ: big five questionnaire. Manuale. Firenze: Organizzazioni Speciali (1993)
26. Penley, J.A., Tomaka, J.: Associations among the big five, emotional responses, and coping with acute stress. Pers. Individ. Differ. **32**(7), 1215–1228 (2002)

The Role of eXtreme Apprenticeship in Enhancing Educational Background Effect on Performance in Programming

Ugo Solitro[1](✉), Margherita Brondino[2], and Margherita Pasini[2](✉)

[1] Department of Computer Science, Università degli Studi di Verona, Verona, Italy
ugo.solitro@univr.it
[2] Department of Human Sciences, Università degli Studi di Verona, Verona, Italy
margherita.pasini@univr.it

Abstract. Informatics is included in school education since many years, but its teaching is developed in many different ways depending on the country, the regional school curricula, and also the local didactic choices. As a result, students start their college studies with very heterogeneous groundings in the subject. Some of these students can encounter significant difficulties in the initial study of programming, especially in non-vocational curricula. In this paper, we consider the first-year students of a degree in Applied Mathematics. We investigate the relations among their background and the difficulties of learning how to program, also connected with the specific teaching methodology.

Keywords: Informatics education · Programming · Learning · eXtreme Apprenticeship · High school college transition · School background

1 Introduction

Informatics permeates almost every aspect of our daily life and it is one of the main factor of innovation and development. In a recent report by Informatics Europe [9] the present situation of informatics education in European countries is described. It is observed that, although Digital Literacy[1] education is in general included in school curricula, Informatics[2] education appears "lacking in most European countries". The risks of inadequate programs for School are highlighted and a number of recommendations have been proposed for the inclusion of Informatics as an autonomous discipline in School curricula. Several papers ([1,4], and others) analyse the specific situation of a country and propose an improvement of school curricula starting from primary level to the upper secondary. A review of models for introducing computer science in K-12 education [6], looking at the situation in different countries, underlines the differences

[1] "Digital Literacy covers fluency with standard software tools and the Internet."
[2] "...the science behind information technology."

R. Gennari et al. (Eds.): MIS4TEL 2019, AISC 1007, pp. 105–112, 2020.
https://doi.org/10.1007/978-3-030-23990-9_13

among educational systems, which makes hard to make a comparison on what is introduced in primary and secondary schools. Many countries make this topics compulsory in primary school and elective in secondary school, and a few countries have made it compulsory in both, while some countries have introduced it only in secondary school. Many countries, among other skills, also introduce Programming languages (e.g. Australia, New Zeland, Norway). It could be interesting to verify whether this programming literacy programs effectively increase academic success in this discipline, in particular programs which introduce to programming languages, compared with students that do not experience these opportunity. At the same time, given that motivation and beliefs could have an impact on performance also in this discipline [7], maybe students' thoughts about what should be thought in a programming course could affect academic performance.

Furthermore, as computational attitudes and skills are regarded as essential for everyone [12], these aspects turn clearly significant for novice students of college science or technical curricula in which advanced computing tools and skills are an essential part of their training. A good approach useful to overcome the early difficulties in programming consists in the adoption of *eXtreme Apprenticeship* (XA, for short). This approach was initially proposed by a group of researchers working in Helsinki (Finland) in a course on programming [10,11] and later proposed again for mathematics [5]. XA is also applied in Bolzano (Italy) for teaching different subjects in a computer science degree (see [2] and [3]).

In the XA methodology, students are stimulated to solve a significant amount of exercises of increasing difficulty; they can get help and suggestions from the support team; but tutors and instructors cannot interfere directly with the solution process.

In this study, we evaluate programming ability in the first step of an academic non-vocational course. This ability was also analysed in connection with the educational background in the high school and with the teaching methodology during the academic program. The main aims of our study, considering the performance on a partial test during the first part of the programming course, were to verify whether:

- the educational background has an effect on performance;
- students' thoughts about what should be the contents in a programming course have an effect on performance;
- the eXtreme Apprenticeship teaching methodology has an effect on performance, as already described in the literature;
- eXtreme Apprenticeship teaching methodology moderates the effect of the educational background.

2 Method

Participants and Description of the Course. The sample included 60 students (60% males, mean age 20.4) enrolled at the first year of the bachelor's degree in Applied Mathematics in Verona. Informed consent forms described the potential participants of the goals of the study and that they could stop their participation at any time during the study. The original sample was larger, but the final sample considered only the sub-group of students for which all the variables of interest could be assessed, given the fact that some of them were not mandatory.

Concerning students' cultural background, 42 (70%) comes from a high school which emphasises technical and scientific subjects; 5 (8.3%) comes from a high school which emphasises humanities, language and social sciences; 9 (15%) comes from an economic-business high school, and 4 (6.7%) from other kind of schools.

The main aim of the course we analyse is the introduction to programming and, in particular, the design and analysis of algorithms using a specific programming language. The course is structured in three periods. The first period is focused on the introduction to programming basics; the second one is mainly devoted to design of data structure; and the major subject of the final period is the analysis of algorithms and data structures. Learning to programming is supported by a number of exercises of increasing difficulties in a lightened XA style. Students are in general required to encode the solution of the exercises; a selection of the exercises, explicitly conceived to sum up a specific subject, are required for submission and evaluation through the Moodle platform. During this first period, the support was provided by a small number of tutors following the guidelines of XA in a lightened form; the assistance is available directly in the programming laboratory and through the Moodle area connected to the course.

The learning progression during the first period is observed by means of a number of tests.

An Entry Test (hereinafter referred to as "**TEST 0**") with no effect on the final grade. It consists in four exercises (with no prerequisites): reading comprehension; informal algorithm comprehension; insertion of an informal instruction in an informal algorithm; detection of errors in an (informal) algorithm.

A Partial Examination performed at the end of the first period. It is divided into in two parts: basic abstract concepts about algorithms and programming, hereinafter referred to as "**TH**"; practical written exercises on code comprehension and programming, hereinafter referred to as "**PR**". A global score, joining together these two parts in a weighted way, is called "**TOT**".

The XA Practice consisting in problem solving and coding developed in the computer laboratory; four distinct activities are submitted and evaluated.

Material and Procedure. In order to preliminarily evaluate students' about previous skills connected with programming and acquired during high school programme, we asked them to fill out a pre-course questionnaire to collect information about their school experience in informatics. This questionnaire asked students the following questions about their high school programme:

1. whether informatics was a subject treated autonomously;
2. if not, which was the subject in which informatics was taught;
3. how many hours they spent on informatics each week;
4. to evaluate on a scale from 0 to 3 their knowledge about some technical terms used in informatics and their knowledge of specific language programmes;
5. students' thoughts about what should be taught in a programming course.

This questionnaire was filled by the students during the first days. Using some of the answers to the pre-course questionnaire, two measures have been defined, as indicators of 1. Knowledge of Programming Languages (**KPL**) and 2. Knowledge of Informatics Terminology (**KIT**). KPL consisted in the sum of the self-report evaluation on 10 specific languages (`markup`, `Pascal`, `Python`, `C`, `C++` or `C#`, `Java`, `JavaScript`, `Basic`, `Ruby`, `Swift`); this variable scored from 0 to 30. KIT was the sum of the self-evaluated knowledge on 4 example of informatics terms (Algorithm, Specification, Programming Language, Markup Language); this variable scored from 0 to 12. We checked the reliability of the two measures, KIT and KPL with *Cronbach's Alpha* and both showed a good index, respectively .77 and .82.

At the beginning of the course, also students' starting knowledge on programming were assessed (TEST 0).

XA practice produced a score evaluating the specific activities. Students were assigned to some programming tasks, under the tutors' supervision. At the end of each group of this programming activity, students should deliver an assignment on this specific activity, for a total number of four deliveries. Evaluation of each assignment was based on taking into account completeness and correctness; a missed delivery was evaluated as zero points. Mean evaluation on the four assignments, joint with a qualitative evaluation of the student-tutor interaction, was a measure of the "compliance" to the XA teaching methodology (XA score). This evaluation has been normalized in the range 0–1.

Performance was assessed considering the partial examination described earlier. In the evaluation the following parameters were considered: correctness of the solution, logical structure and good programming practices. The test was composed of two parts: a general "theoretical" section, in which the knowledge of fundamental notions (e.g. the definitions of compiler, interpreter, specification) are verified; a practical "programming" section, where the student must solve a few exercises of increasing difficulty about programming competences and problem solving skills. The evaluation in this test produced a quantitative score, which was normalized in the range 0–1. At the end, "performance" were measured with three different-even if related-quantitative dependent variables: total score (TOT), theoretical score (TH), and programming Score (PR).

In summary, the measures considered in this study are the following:

1. **KPL**: Knowledge of Programming Languages;
2. **KIT**: Knowledge of Informatics Terminology;
3. **XA**: compliance with the eXtreme Apprenticeship methodology;
4. **TEST 0**: starting knowledge in programming;
5. **TOT**: performance in the partial examination (global);
6. **TH**: Theoretical score in the partial examination;
7. **PR**: programming score in the partial examination.

In addition to these measures, the answers to the other questions in the pre-course questionnaire have been considered.

Data Analyses. First, descriptive statistics were computed, to sum up the level of the considered variables. In a second step, independent sample t-test was used to verify whether the performance was different for students who did or did not do a certain activity during the high school. Finally, hierarchical regression models were used to verify whether KPL, KIT and XA predict performance (in terms of global performance, theoretical score and programming score, and to test the hypothesis that XA moderate the effect of KPL and KIT on performance.

3 Results and Discussion

Educational Background: Descriptive Statistics and Its Effect on Performance. For 23 students (38.3%) of informatics was a subject treated autonomously and for 2 (3.3%) it was part of the program of another subject (mainly Mathematics, Physics and Sciences). These students generally studied informatics for two hours a week; for 16 students (30%) this subject was treated only marginally or, and for 19 students (31.7%) this subject was not at all part of the program. Considering two groups of students, one joining together the students who had informatics as an autonomous subject or as part of the program of another subject, and the second one joining together the students who treated the subject marginally or not at all during their high school, we could perform some independent sample t-test using the performance as the dependent variable. For all the three variables (TOT, TH and PR) students of the first group performed significantly better than students of the second group, with a large effect size. Table 1 shows these results.

Table 1. Mean, standard deviation, t, p-value and effect size (*Cohen's d*) comparing the two group of students in performance.

Performance	Autonomously treated		Marginally treated		t(52)	P	Cohen's d
	M	sd	M	sd			
TOT	0.73	0.22	0.58	0.19	2.66	0.01	0.73
TH	0.7	0.22	0.58	0.19	2.27	0.027	0.62
PR	0.54	0.19	0.42	0.18	2.54	0.018	0.67

During informatics lessons at the secondary school, 93% declared to use web browsers, 78% spreadsheets, 77% a word processor, 57% a mathematical software, 40% used programming languages, 23% used instruments to develop web pages, and 10% instruments for Integrated Development Environment (IDE).

Only students who used mathematical software and programming languages showed a better performance compared to the group of students who did not use them (see Table 2). IDE was not considered because of the small number of students, and also because it co-varies with the other two.

Table 2. Mean, standard deviation, t, p-value and effect size (*Cohen's d*) comparing the two group of students in performance.

Studied at school	Performance	No		Yes		t(52)	P(2 tails)	Cohen's d
		M	sd	M	sd			
Mathematical software	**TOT**	0.55	0.17	0.71	0.22	−2.82	0.007	−0.78
	TH	0.58	0.16	0.67	0.23	−1.67	0.101	−0.46
	PR	0.38	0.17	0.54	0.18	−3.17	0.003	−0.88
Programming languages	**TOT**	0.57	0.20	0.75	0.20	−3.22	0.002	−0.89
	TH	0.57	0.20	0.71	0.20	−2.66	0.010	−0.73
	PR	0.41	0.18	0.56	.17	−3.00	0.004	−0.83

Students' thoughts about what should be taught in a programming course didn't affect performance: no t-test highlight significant differences in the three kind of performance comparing the group that think a specific thing with the group that does not think it. The majority of the students think that the programming course should teach how to solve problems using computing instruments (83%), how to use a computer (78%) and how to use specific software (77%). 40% think that the course should teach the use of programming languages. Only few students think that the course should teach website design, or ICT.

Considering Knowledge of Programming Languages and Knowledge of Informatics Terminology, these two indicators showed a positive correlation with both the first assessment of the starting knowledge in programming and with the three kind of performance, as shown in Table 3. In general, knowledge of informatics terminology showed a larger relationship.

Educational Background and XA for the Prediction of Performance. Results described until now seem to suggest that educational background, in terms of knowledge of programming language and knowledge of informatics terminology, has an effect on performance. But what is the role of teaching methodology? We know from the literature that XA affect performance. This second step of the analysis aims to understand whether there is a conjoint effect of the two aspects – educational background and XA – in affecting performance in programming. To do that we run some hierarchical regression analyses, inserting in a first step KIT/KPL and XA, and in a second step the interaction term (KIT × XA or KPT × XA) to explore the moderation effect of XA. Results showed that the moderation effect is present mainly for KPL: KPL does not predict performance (nor for TOT neither for TH and PR), but the insertion of XA in the model, and of the interaction term, showed that students XA – that has a high predicted power itself on performance – enhance the role of KPL, as shown in Table 1. High level of XA predict a better performance (TOT) in general, but for students with a higher level of KPL the effect of XA was higher. The interaction effect was significant considering an *alpha* of .1 (p = .073) and at this step this has been considered acceptable, also given the small sample size. The effect is in the similar direction also for TH and for PR. The situation is different when considering KIT. Knowledge of informatics terms has an effect on performance (TOT: p = .019; TH: p = .029; PR: p = .068), and no inter-action effect with XA was found.

Table 3. Pearson's correlations between the two indicators of programming competences at the High School level and performance (both at TEST 0 and on the other tests).

	TEST 0	**TOT**	**TH**	**PR**
KPL	.38**	.31*	.25	.29*
KIT	.54**	.51**	.45**	.47**

*p < .05; **p < .01

4 Conclusions

This study aims to understand which aspect of the educational background concerning informatics could help students in their approach to programming, mainly in a non-vocational course. In a previous research [8] we analysed the impact of different educational experiences on students' performance in developing programming abilities, only considering the kind of school attended by the students, highlighting that human-oriented learners showed more difficulties than sciences-oriented colleagues, more in practical tasks than in theoretical ones. In the present research, the further step was to try to have a metric measuring the self-report knowledge evaluation on 10 specific languages and on 4 example of informatics terms. A second aim is to confirm the important role of teaching methodology, in the present case XA, also in enhancing the role of other basic requirements acquired in the high school experience. Results confirm the importance of students' exposition to some basic concepts concerning informatics, in general the ones related to Computational Thinking, for a better performance in programming. Coding practice and the knowledge of the informatics terminology, together with a minimal knowledge of a programming language, could be the first step for an effective learning in a programming course. Along with this, the use of an appropriate teaching methodology, as XA, can ensure the best results, enhancing the role of these previous competences. More research is needed to better understand the mechanism, and to understand whether the absence of the moderation effect in some cases is due to sample size problem or reflect the real situation.

References

1. Barendsen, E., Grgurina, N., Tolboom, J.: A new informatics curriculum for secondary education in The Netherlands. In: Lecture Notes in Computer Science including subseries Lecture Notes in Artificial Intelligence and Lecture Notes in Bioinformatics (2016)
2. Del Fatto, V., Dodero, G., Gennari, R.: Assessing student perception of extreme apprenticeship for operating systems. In: 2014 IEEE 14th International Conference on Advanced Learning Technologies, pp. 459–460. IEEE (2014)
3. Del Fatto, V., Dodero, G., Lena, R.: Experiencing a new method in teaching Databases using Blended eXtreme Apprenticeship. Technical report, DMS (2015)
4. Forlizzi, L., Lodi, M., Lonati, V., Mirolo, C., Monga, M., Montresor, A., Morpurgo, A., Nardelli, E.: A core informatics curriculum for italian compulsory education. In: Informatics in Schools. Fundamentals of Computer Science and Software Engineering, pp. 141–153 (2018)
5. Hautala, T., Romu, T., Rämö, J., Vikberg, T.: Extreme apprenticeship method in teaching university-level mathematics. In: Proceedings of 12th International Congress on Mathematical Education, ICME (2012)
6. Heintz, F., Mannila, L., Färnqvist, T.: A review of models for introducing computational thinking, computer science and computing in k-12 education. In: 2016 IEEE Frontiers in Education Conference (FIE), pp. 1–9. IEEE (2016)
7. Shell, D.F., Hazley, M.P., Soh, L.-K., Ingraham, E., Ramsay, S.: Associations of students' creativity, motivation, and self-regulation with learning and achievement in college computer science courses. In: 2013 IEEE Frontiers in Education Conference (FIE), pp. 1637–1643. IEEE (2013)
8. Solitro, U., Zorzi, M., Pasini, M., Brondino, M.: Computational thinking: high school training and academic education. In: GOODTECHS Conference Proceedings (2016)
9. Vahrenhold, J., Nardelli, E., Pereira, C., Berry, G., Caspersen, M.E., Gal-Ezer, J., Kölling, M., McGettrick, A., Westermeier, M.: Informatics education in Europe: are we all in the same boat? Technical report, Informatics Europe, May 2017
10. Vihavainen, A., Luukkainen, M.: Results from a three-year transition to the extreme apprenticeship method. In: Proceedings of 2013 IEEE 13th International Conference on Advanced Learning Technologies ICALT 2013, pp. 336–340 (2013)
11. Vihavainen, A., Paksula, M., Luukkainen, M.: Extreme apprenticeship method in teaching programming for beginners. In: Proceedings of 42nd ACM Technical Symposium on Computer Science Education - SIGCSE 2011, p. 93 (2011)
12. Wing, J.M.: Computational thinking. Commun. ACM **49**(3), 33–35 (2006)

Grab that Screen! Architecture of a System that Changes the Lecture Recording and the Note Taking Processes

Marco Ronchetti(✉) and Tiziano Lattisi

DISI, Università di Trento, via Sommarive 9, 38123 Povo, TN, Italy
marco.ronchetti@unitn.it

Abstract. We present an innovative system which couples the ability of producing recordings of lectures, meetings and other events when something is presented via screen projection, with the possibility of taking screenshots of what is being projected at any give time. This opens new possibilities for more efficient note taking, which is a practice known to trigger important cognitive activities.

Keywords: Video lectures · Note taking

1 Introduction

Technology impacts on our behavior, as it offers new and often more efficient ways to perform operations. The last decade has seen the wide diffusion of smartphones, which like (and probably more) than other technologies modified our habits. One example is the grabbing of images: very often and for various reasons we capture what we see. Conferences, seminars and classes are no exception. Quite frequently we can see students recording the audio of a lecture by using their phones, or we can notice conference delegates taking pictures of the screen during a presentation. Not always however the results match the expectations. Pictures can be blurred, and recordings can be noisy. But even in the lucky case, when the result is technically sound, there is still a problem: the context is missing. So we may end up with pictures, which contain a readable content, but we do not remember exactly about what it is. Moreover, these multimedia resources are decoupled from the notes we are taking.

We present here the architecture of a system, which solves this problem, because it creates an infrastructure, which enables us to capture multimedia resources integrating them into our own notes. This can be easily done in classroom, in a conference, at a meeting: whenever anything is projected on a screen to show us something. What we capture can be integrated with any tool on a laptop (office application, mind maps, editors of various kind, on-line tools), or directly kept, annotated and maintained in a web application. As such, it has direct applications in academic and scholar contexts.

In this paper, we will first discuss why note-taking is believed to be important and then we proceed to present the architecture we propose. Finally we discuss the results and draw our conclusion.

R. Gennari et al. (Eds.): MIS4TEL 2019, AISC 1007, pp. 113–120, 2020.
https://doi.org/10.1007/978-3-030-23990-9_14

2 Grounding Our Research

According to many scholars (see e.g. [1, 2]), note-taking is an important cognitive activity. It comprises two phases: note taking and note review. During the first phase (e.g. in class), the student activates an encoding process. This forces her/him to keep a high level of attention, organize ideas, and link the presented material to her/his existing knowledge. In the second phase, the student retrieves the information, which might have been forgotten in the meantime, reviews it and consolidates the acquired knowledge.

The positive effect of note-taking is confirmed by a meta-analysis by Kobayashi [3]. Some people disagree (see e.g. [4]), arguing that personal notes lack effectiveness due to their incompleteness, but they seem to focus more on a view of the notes as a replacement of a textbook (and hence on knowledge transmission). In fact, note taking can be seen in two quite different ways: either as a "content capture" process, or as a selective, critical and personal elaboration of what is presented by the on-going activity. The second one is closer to a constructivist approach to learning. By the way, this applies to any learning activity – not necessarily a frontal presentation – but also to other endeavors such as e.g. a discussion, or a meeting.

Interestingly, a recent study indicates that longhand note-taking is better than laptop based note-taking [5], probably because many students are faster at typing than at writing, and hence the ones using laptops are more likely to simply transcribe the teacher saying. This had echo also on the New York Times, which concluded that laptops should be banned from class [6].

The challenge seems hence to be able to relieve the listener from the duty of "capturing content": if it is not necessary to jot what we hear and what we see, we can focus more on "active listening" and take more significant notes. It is in fact quite common, that teacher make copies of the presentations they used for lecturing available before the lecture itself (e.g. by uploading the corresponding PDF files on a leaning management system such as e.g. Moodle). Students can then either print them and take notes over the printouts, or annotate them on their laptop or tablet. In other context such as conferences or company meetings however it is far less common that printouts of the presentation(s) be available. Also lectures, which are not based on presentation software, such as e.g. practical demonstrations, even when they are computer-based, are unlikely to offer any type of transcript.

The availability of printouts is not by itself a solution, and sometimes has important drawbacks. Far too many students, especially among freshmen, tend to believe that they can prepare an exam simply by reading (and possibly memorizing) the content of the available PowerPoint slides. Fjortoft [8] even reports a case, in which students reported that one of the main incentives to be present in class was the unavailability of handouts! Also, since the material is already there, many students do not even feel the need to take notes, aborting hence a cognitive activity that, as we discusses in the opening of this section, is widely believed to have a positive effect on understanding and learning.

A different approach is offered by systems like Classroom Presenter [8], where one of the features allow students to annotate a slide which is provided by the teacher. DyShare [9] also allows students to take notes over what the teacher is (explicitly) providing them.

On the basis of what we discussed so far, we run a preliminary investigation on 100 Computer Science freshmen. 96% declare they take notes during lectures, with 69% doing it only on paper and 28% only digitally. 30% declare they have difficulties in keeping up with their notes during the lecture. 87% would welcome the system we will describe here. Further details about this investigation will be reported elsewhere.

3 System Architecture

The system we developed is based on a low cost hardware device called Lodebox. It is based on a commercial microcomputer having small but sufficient computing power, and offering several I/O ports. In particular, it has HDMI input and output, audio input (also audio output, which is however not used), USB ports, which we use for connecting input devices to perform a minimal set of actions, Ethernet and wireless networks. The Lodebox is located between the video output of a standard PC, and the video input of a common projector. It replicates on the video output port whatever it gets on the video input port, and it is able to perform operations on the flow of data, which passes through it.

3.1 Note-Taking Functionality

One of these operations is to grab an image on demand. Requests for grabbing come (via HTTP) from a Web Application running on a standard Web Server. These requests are triggered from light processes running on client devices, and the response to a client contains the grabbed screenshot. The client's process makes then the obtained resource available to any other application running on the client itself.

From the user point of view, this simply means activating the request (e.g. by clicking on a button) and then telling the application s/he uses for taking notes to use the grabbed screen.

Figure 1 depicts the overall architecture supporting this functionality.

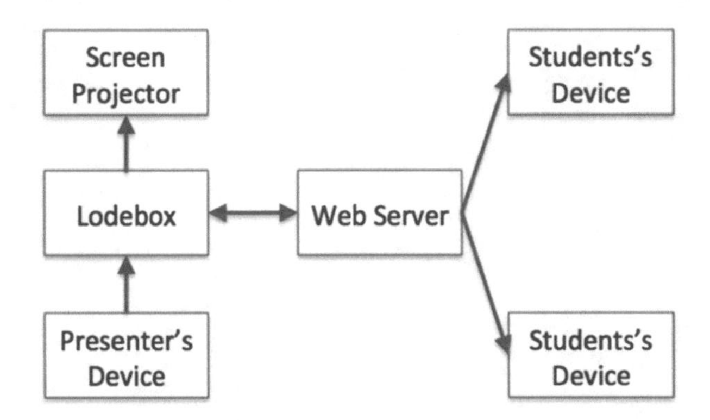

Fig. 1. Basic capture functionality

Figure 2 shows and advanced version of the capture functionality. It adds a database, which allows keeping track of the users and can provide a container where a more complex Web Application running on the server saves the history of screenshots grabbing, and the annotation users add by using the Web App's front end.

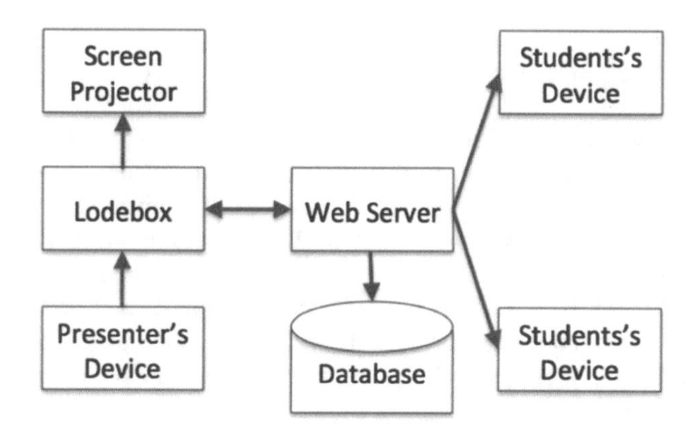

Fig. 2. Advanced capture functionality

An example of such front end is presented in Fig. 3. Here number 1 shows the grabbed screenshot, which is annotated with a colored virtual pen, chosen among the tools offered by the toolbar (2). 3 shows a rich text editor component, and 4 a set of textual notes. The user can navigate through the various screenshots, and also add blank pages to add extra notes.

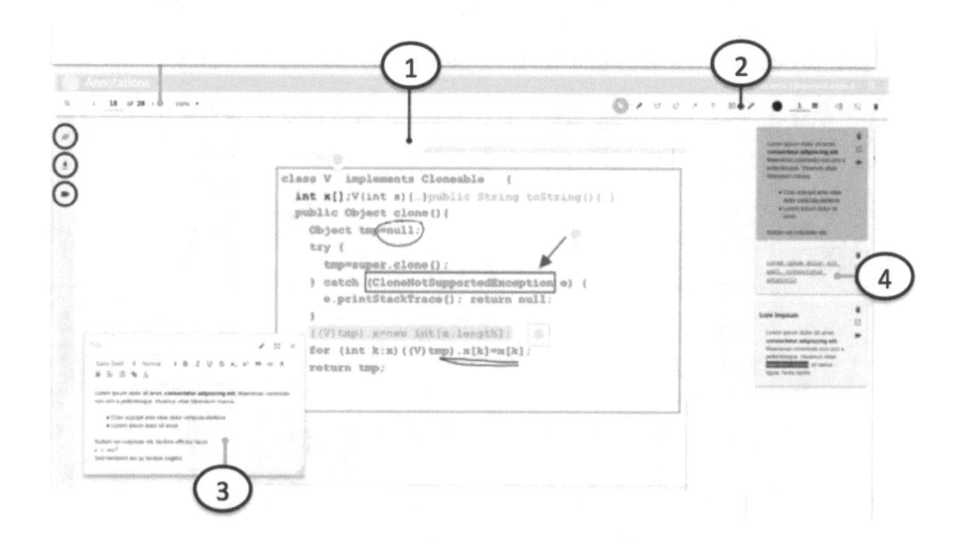

Fig. 3. Example of the Web App front end

Such Web App, which is written is Angular, is a demo of a Rich Interactive Application, which builds on top of the capture functionality offered by the Lodebox. By using it, the user can define contextual units (e.g. a lecture, a meting or a seminar) where her/his own notes are mixed with what s/he captured.

3.2 Recording Functionality

The Lodebox has other capabilities. By using the other input ports (wireless network and audio in) it can simultaneously record three media streams: a video stream coming from a wireless camera, an audio stream coming from a radio microphone and the video stream which traverses it, going from the presenter PC to the projector. These streams are saved on local data storage. Figure 4 presents this scenario.

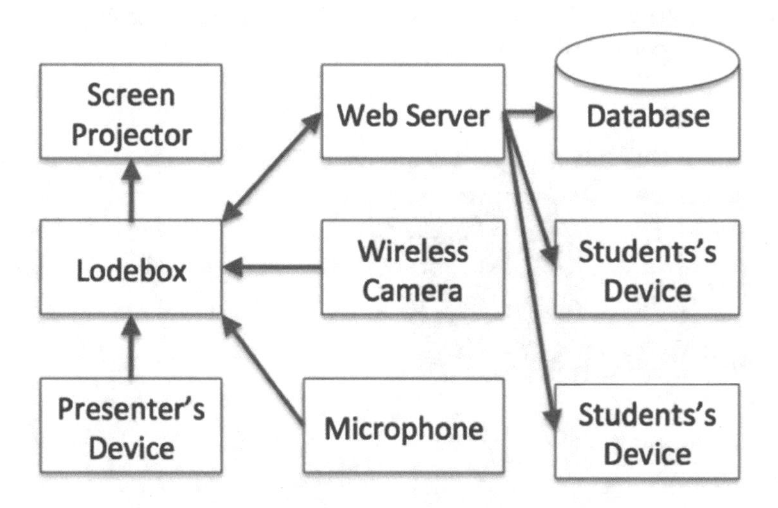

Fig. 4. Three-streams recording functionality

The recording functionality is activated by using a menu, which during the setup phase is shown on the projector, and which can be commanded via a simple device such as a numeric keypad connected via USB, or via pushbuttons integrated in the Lodebox itself (depending on the actual hardware implementation).

After the end of the recording, the collected data are transferred to a machine (a PC), where a simple user interface allows performing post-processing, which consists in compacting and merging the media stream, and generating a wrapper. A stand alone HTML5 is hence produced, which can be published on any web server to show the recorded video lecture. The result is somehow similar to the one provided by other screen recording systems, but with some differences: in first place here we are not limited to static images, but rather two synchronized video streams are presented at the same time. The user has the option to view them at equal size, or put the cognitive focus on one of them making it big, while the other video stream becomes small. The operations necessary to capture the lecture are fully transparent to the speaker (who does not need to bother with the infrastructure) and are made fully automatic, up to the loading of the resulting video-lecture on a server. Finally, the proposed software and hardware solution is much cheaper than most of the currently available video-recording systems.

3.3 Putting It All Together

A big advantage finally comes when the two functionalities are put together. In fact, the generated screenshots that the user can capture contain some extra information about the context. These information can integrate the notes and the video recording in yet another web application. By using this client through a standard HTML5 - compliant browser, the user can start from his/her notes to review what happened at the time when the note was taken. The notes can be then seen as a semantic index of the video-lecture. In a dual way, the video-lecture becomes supporting material for the notes, as the user can retrieve the context when the note was taken.

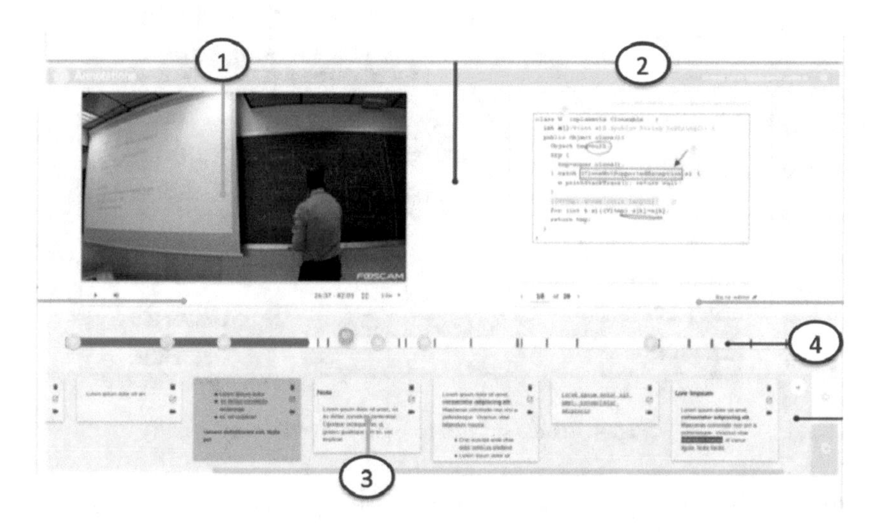

Fig. 5. Annotated video-lecture

Figure 5 shows a prototype client for this functionality. Numbers 1 and 2 show the two video streams: the one captured from the camera, and the one showing whatever was projected on the screen. Annotations taken by the user using the client we discussed in Fig. 3 appear on top of the flow n.2. The textual notes are shown in region 3: by clicking on one of them, the two video stream resynchronize at the time when the note was taken. A time bar (4) shows markers where notes were taken, and can be used to navigate the lecture.

4 Discussion and Conclusions

In this paper we presented a software + hardware architecture that allows an innovative functionality: capturing screen shots of whatever is being shown on the projected screen in a classroom, a conference room or a meeting room. The system has been actually implemented and successfully tested. A similar functionality can be found in literature in the work by Anderson et al. [8], but is limited to capture slides explicitly shared by the teacher. DyKnow, by Berque [9] allows students to capture on an electronic notebook what the teacher shares: the teacher needs to import the material s/he wants to share in an ad-hoc sharing software. Both approaches require an intentionality by the teacher. Our system instead allows students to capture and annotate just whatever happens to be projected by the presenter, without any intentionality by the teacher, without requiring him/her to use any special software, and without any limitation to specific artifacts to be shares, such as e.g. slides.

As far as we could search, we found no system offering anything actually similar to what Lodebox offers.

Moreover the Lodebox system also allows recording an event (lecture, conference or meeting) capturing an audio stream and two video streams (a camera and whatever is being projected on the screen: slides, but also live demonstrations, videos, internet browsers, etc.). The recorded streams are then, through automatic post-processing, published on a web site. Although many other systems can produce video-lectures, the Lodebox offers some features – such as a synchronized double stream – which are not commonly offered. What makes the proposed system unique, however, is its capability of synchronizing the notes that the user takes “in vivo” with the video-lecture itself.

So far we could only test the system with a limited number of users in a laboratory environment, but we plan for an extensive validation study soon.

References

1. Di Vesta, F., Gray, S.: Listening and note taking. J. Educ. Psychol. **63**(1), 8–14 (1972)
2. Peper, R.J., Mayer, R.E.: Generative effects of note-taking during science lectures. J. Educ. Psychol. **78**(1), 34–38 (1986)
3. Kobayashi, K.: Combined effects of note-taking/-reviewing on learning and the enhancement through interventions: a meta-analytic review. Educ. Psychol. **26**(3), 459–477 (2006)
4. Kiewra, K.A.: Providing the instructor’s notes: an effective addition to student notetaking. Educ. Psychol. **20**(1), 33–39 (1985)
5. Mueller, P.A., Oppenheimer, D.M.: The pen is mightier than the keyboard: advantages of longhand over laptop note taking. Psychol. Sci. **25**(6), 1159–1168 (2014)
6. Dynarski, S.: Take Notes With Pen and Paper? It Can Be Done. The New York Times, 26 November 2017, on Page BU4. Online version: Laptops Are Great. But Not During a Lecture or a Meeting. https://www.nytimes.com/2017/11/22/business/laptops-not-during-lecture-or-meeting.html?emc=eta1. Accessed 26 Nov 2017
7. Fjortoft, N.: Students’ motivations for class attendance. Am. J. Pharm. Educ. **69**, 107–112 (2005)

8. Anderson, R., Anderson, R., Chung, O., Davis, K.M., Davis, P., Prince, C., Razmov, V., Simon, B.: Classroom presenter: a classroom interaction system for active and collaborative learning. In: WIPTE (2006)
9. Berque, D.: An evaluation of a broad deployment of DyKnow software to support note taking and interaction using pen-based computers. In: CCSC: Northeastern Conference (2006)

Designing a Self-regulated Online Learning Course Using Innovative Methods: A Case Study

Leonardo Caporarello[1], Federica Cirulli[2(✉)], and Beatrice Manzoni[1]

[1] SDA Bocconi School of Management, Bocconi University, Milan, Italy
{leonardo.caporarello,beatrice.manzoni}@unibocconi.it
[2] Bocconi University, Milan, Italy
federica.cirulli@unibocconi.it

Abstract. Students from Gen Z ask for being independent as well as guided by the adults in their learning experience, especially when learning is aimed at making better future career and development choices. Self-regulated online learning can effectively answer this need. While its definition, advantages and phases have been debated in the literature, self-regulated online learning still lacks an in-depth analysis. In this paper, we adopt an action research approach to provide a design framework for an online course specifically oriented at supporting students in their job-related choices. From a theoretical point of view, we use a framework based on the Pintrich's self-regulated learning approach. From a practice point of view, we provide a set of recommendations to design effective self-regulated online learning experiences.

Keywords:: Self-regulated learning · Online learning · Gen Z · Action research

1 Introduction

Current students, belonging to the "Generation Z" (born from 1997 onwards), expect a learning experience which is permeated by new technologies [1]. They are the first real "digital natives", given the extent to which technology has been part of their lives since their early years [2]. The interest towards them is growing because they are ready to enter the job market and many organizations are wondering how to accelerate their onboarding and learning processes within the organizational context [2].

When it comes to learning, individuals belonging to this generation want to play an active role in their learning, even more active compared to other previous generations. They want to get involved in the process which takes place in and outside class, and to learn experientially [3]. Moreover, they want to be empowered, autonomous and independent in organizing the learning resources made available to them, even if at the same time they need a feedback and a strong guidance from adults in selecting what is really relevant to them [4]. The pervasive technology that is part of their life facilitates and supports the learning experience, but it complicates it as well, giving access to infinite learning options and learning resources and making any decision hard [3, 5].

R. Gennari et al. (Eds.): MIS4TEL 2019, AISC 1007, pp. 121–128, 2020.
https://doi.org/10.1007/978-3-030-23990-9_15

In this situation, universities can play a very important role in designing learning experiences that help Gen Z to select relevant information [1], acquire knowledge and skills and increase their self-efficacy, in order to be ready to face the job market [6].

Given this, how should universities design a self-regulated online learning experience in order to help Gen Z make decisions with regard to their future career?

To answer the research question, we apply and expand Pintrich's self-regulated learning design framework. At the time of writing, we are designing an online course that helps students make decisions about their career before approaching the job market.

2 Self-regulated Online Learning

Over the last two decades, the interest towards self-regulated learning increased, with regard to its meaning [7, 19] and its phases [8–16].

However, there is a lack of standard and aligned definitions [17]. Most scholars consider self-regulated learning as "the self-directive processes and self-beliefs that enable learners to transform their mental abilities, such as verbal aptitude, into performance skill, such as setting goals" [18 p. 166]. It is useful to support students in developing their capacity of judgment [21] and it requires that students' choices are challenging and feasible at the same time. Self-regulated learning is especially important when taking online courses [22, 23] as online learning environments are characterized by a high degree of autonomy [24]. Learners are in charge of their own learning: they set personal target goals and regulate and control many aspects influencing their learning [25].

In such perspective, self-regulated learning has many advantages such as a generally positive impact on the academic results and on the development of critical judgment skills [26], as well as on the emotional involvement and motivation towards the course [21].

In order to define an effective self-regulated learning process, several authors explored its main phases [8–15]. For example, Zimmerman [9] studied three phases: the forethought phase, the performance phase and the self-reflection phase. Boekaerts [8] illustrated the concept of learning episodes, where learners are asked to activate decision making processes demonstrating context-specific and goal-directed learning behaviour [8]. According to these studies, self-regulation learning is viewed as an activity that learners do for themselves in a proactive manner rather than something that happens to them in reaction to teaching.

Due to advancements in technology, self-regulated learning increasingly occurs in an online or blended environment [10, 11]. Some scholars studied the impact of technology on self-regulated learning, suggesting a positive impact in terms of students' increased sense of ownership and control over their learning [12]. Self-regulated learning via mobile devices also allowed for a better student-centred approach and a ubiquitous learning [13]. Other studies stated that it is extremely relevant for designers and instructors to know the phases of self-regulated learning and how to turn learning environments into "energizers" for students [14].

In such regard, Pintrich [15] elaborated a framework including four phases, which is still the most frequently used one [27]. The first stage is named "forethought, planning and activation" and it involves planning, goal setting and activation of students' interest about the contents and the context in which they will learn. The second stage – "monitoring" – implies acquiring capacity of judgment and self-awareness about the content. The third stage – "control" – involves efforts to control and regulate different aspects of the self or task and the context. The last and fourth stage – "regulation" – refers to various reactions and reflections of the self in relation to tasks or context.

Despite being among the most cited frameworks in the literature about self-regulated learning, Pintrich's framework still lacks empirical testing, in particular within an online or blended setting. Moreover, despite its appropriateness for improving decision making skills, its use in this sense is still unexplored [28]. In such perspective, through action research, we aim at developing an integrated version of Pintrich' framework, providing design guidelines.

3 Research Method

We carry on an action research project with the Career Service - Markets and External Affairs Division of Bocconi University in Milan (Italy). The Career Service supports both students looking for the best-suited jobs and companies looking for the right candidates. They help an ambitious and confused generation of future professionals in the delicate moment where they are starting to approach the job market. We are working with the Career Service to design and implement an online course – named "Career Service Starter Pack" – that aims to prepare and orient students in the job market. The project team is multidisciplinary, and involves content experts, graphic designers, communication specialists, instructional designers, and a project manager.

The research method used for the scope of this paper is the action research, which aims at producing practical knowledge that is useful to people [29], through a continuous cycle of developing and elaborating theory from practice [30] and a consensus-building process that involved in our case the Career Service and the design team [31]. Subjects and researchers are jointly responsible for developing and evaluating theory to ensure that the results of the research help solve a real challenge and reflect the knowledge created through the participative process.

In our case, the real challenge is the development of an online course that support students in making career related decisions. While designing the course for the Career Service team, we will also expand the theory on self-regulated online learning.

Broadly speaking, our research entails phases of groundwork, intervention, theory testing and development.

Although presented sequential, the phases were iterative and cyclical. In the *groundwork* phase we sought to understand the need of the Career Service, the target, the other existing services already provided. In parallel, we also search for external case histories and best practices, even conducting a literature review. We developed a working agreement that clarifies mutual expectations between us and the Career Service staff.

Then, in the *intervention* phase, we met with the Career Service team once a week to design and micro-design each module. Periodic interventions have enabled the team to articulate and question their proposals and apply them in practice. We often systematized interactions, providing a toolbox of questions that helped the team to more deeply examine the issues. The process has been participatory, collaborative and reflexive. In each meeting, team members were invited to reflect about learning outcomes, contents, and structure of each module. The suggestions collected were discussed; if not accepted, we modified learning outcomes, refined contents, and adjusted the module structure.

In the *theory testing and development* phase we sought to formulate, evaluate and revise our understandings of the online self-regulated learning process within coherent theories. This phase closes the "circle" of action research. In our case, the theory testing phase is in progress. Actually, we are analysing preliminary qualitative feedbacks about the course effectiveness. Based on students' preliminary feedbacks we are reflecting on how the course can be revised and improved.

4 Findings from Designing the Career Service Starter Pack

The "Career Service Starter Pack" is a self-regulated online course, offered to students that are at the end of their university studies, in order to facilitate their orientation in the job market. The efforts in the design phases aim at creating an enhanced and unique learning experience. Course's contents have been chosen after a benchmarking of the existing European courses on job market orientation, in order to select the most important topics in this domain. The course is available through LMS, and it is complemented with additional offline orientation activities on campus. It includes six modules on different topics (see Table 1). The learning journey is modular and personalized, in the sense that students can autonomously choose the modules they want to attend and thus build their own course.

Considering the design of the course, we use an adapted version of Pintrich's self-regulated learning framework.

With regard to the "forethought, planning and activation" phase, we develop a visual infographic, which sum up the main learning outcomes and contents of each module and focuses learners' attention on the topic. Learning outcomes emphasize the application and integration of knowledge, so they have been accurately selected and described in the infographic. Furthermore, the effectiveness of this phase is enhanced involving all the team, with weekly meetings, in the choice of the more suitable topic for each module in relation to specific learning outcome, in the selection of each content in order to support decision making, and in the selection of the more appropriate method to deliver each content. This online section is also reinforced by a subsequent face-to-face individual meeting with a Career Service member, to which the student can arrive more prepared and self-conscious than in the past. In fact, students can benefit of face-to-face activities that have already planned within the Career Service (e.g. meetings on how to write the CV and the cover letter, how to approach a job interview or a group assessment; meetings on personal branding strategies and the use of social media). These sessions represent the occasion for a deep-dive into specific

Table 1. Modules structure and Learning outcomes

Modules	Learning objectives
Personal branding and social reputation	Know the fundamental elements of storytelling and understand how to use it during an interview Understand social reputation strategies, know how to recognize them and eventually apply them
CV	Know the basic rules for an effective CV Know what recruiters are looking for Know how to draw up different CV for different countries: global perspective
Cover letter	Know the elements of the cover letter Know how to draw it up
Applying online	Understand the characteristics of internships/job online applications Know how to apply online
Gaming	Understand the role of gaming in employers' attraction and selection process Be more aware of what recruiters expect and analyse in attraction or selection game
Networking	Know and apply essential actions aimed at building, expanding and maintaining professional relationships Know what an elevator pitch is and how to prepare for one

topics, which have been already explored online. The online resources are delivered to present the learning contents before these meetings, where students are capable to interact with their peers and staff for in-depth discussions. Monthly, students are invited to read online contents through e-mails, social medias, journal articles, learning environment announcements. This is in line with a self-regulated learning which is "highly context dependent" [32]: students are considered as not isolated but an integral part of the institution. This form of "engagement" can lead to a more effective acquisition of knowledge.

The "monitoring" phase is planned to check the students who attend the online course and verify that they are the same who attend the face-to-face meetings in order to ponder the effective consolidation of knowledge. For this scope, interviews with students are also conducted. Interviews are not included in the traditional Pintrich's framework but, if introduced, they can be a valuable source of feedbacks and ongoing learning for both learners and instructors.

Then, the "control" phase involves efforts to control and regulate different aspects of the self or task and context. Some tasks can be solved in the course, for example how to write a CV or how to network for a professional purpose; others will be discussed in face-to-face meetings. These moments are organized to explain unclear issues of the tasks as well as a students' opportunity for sustained exploration of most difficult topics.

Finally, the "regulation" phase includes reaction and reflections on the self and the task or context. For this reason, the online course ends with a "How to" downloadable guide, summarising the key takeaways for each module and listing a checklist of actions the student should perform to be ready for the job market. The efficacy of this phase is enhanced thanks to the individual meeting that Career Service offers also to discuss about unclear topics.

5 Discussion and Conclusion

In this article, we provide a research as well as a practice-oriented contribution in the stream of self-regulated online learning.

From a research point of view, we propose a design based on Pintrich's self-regulated learning framework. Several scholars have discussed the phases of a self-regulated learning experience, but few studies have delved into their suitable application online [33]. Yet, this is critical also considering the fact that the new generations are digital natives who live and learn in a tech-immersive context.

From a practice point of view, we recommend the use of blended learning models, instead of just online or face to face ones. In our design, Pintrich's activation phase is more effective thanks to the combination of online experiences and face to face meetings with the Career Service staff. Moreover, we suggest to both monitoring and providing feedbacks to students related to their online activities and accomplishments of the tasks. Finally, we suggest supporting the learning process with a formal recap and debriefing of the main key concepts.

In terms of directions for future research, we plan to measure the perceived effectiveness of the course, that means also of its design using the Pintrich's framework.

This feedbacks analysis, which is not included in the original framework, represents a value added from a research and practice point of view. More specifically, and in line with existing research, we will measure performance expectancy [34], effort expectancy [34], social influence [34], facilitating conditions [34], attitude toward it [34], intention to use [34], personal innovativeness [35] and the student's preferred learning style [36].

At this stage of the research, we collected preliminary qualitative feedbacks on the course from the students who used it so far. Students appreciate the contents that are exhaustive and hands on, but also the clarity in terms of learning flow and the easiness to use. Areas of improvement relate to the content and in particular to the need of adding industry specific insights. A quantitative survey using validated scales is ongoing. Final results will be available over the next months, allowing for providing further recommendations for instructional designers.

References

1. Rashid, T., Asghar, H.M.: Technology use, self-directed learning, student engagement and academic performance: examining the interrelations. Comput. Hum. Behav. **63**, 604–612 (2016)
2. Mohr, K.A.J., Mohr, E.S.: Understanding Generation Z students to promote a contemporary learning environment. J. Empower. Teach. Excel. **1**(1), 9 (2017)

3. Morton, C.E., Saleh, S.N., Smith, S.F., Hemani, A., Ameen, A., Bennie, T.D., Toro-Troconis, M.: Blended learning: how can we optimise undergraduate student engagement? BMC Med. Educ. **16**(1), 195 (2016)
4. Azzam, A.M.: Why students drop out. Educ. Lead. **64**(7), 91–93 (2007)
5. Van der Kleija, F., Vermeulena, J.A., Schildkampb, K., Eggen, T.J.H.M.: Assessment in education. Princ. Policy Pract. **22**(3), 324–343 (2015)
6. Sun, Z., Xie, K., Anderman, L.: The role of self-regulated learning in students' success in flipped undergraduate math courses. Internet High. Educ. **36**, 41–53 (2017)
7. Verdín, D., Godwin, A.: First in the family: a comparison of first-generation and non-first-generation engineering college students. In: Proceedings of 2015 IEEE Frontiers in Education Conference, pp. 1–8 (2018)
8. Boekaerts, M., Cascallar, E.: How far have we moved toward the integration of theory and practice in self-regulation? Educ. Psychol. Rev. **18**, 199–210 (2006)
9. Zimmerman, B.J.: Becoming a self-regulated learner: an overview. Theory Pract. **41**(2), 64–70 (2002)
10. Woolfolk, A.E., Winne, P.H., Perry, N.E.: Educational Psychology. Allyn and Bacon, Scaborough (2000)
11. Boekaerts, M., Corno, L.: Self-regulation in the classroom: a perspective on assessment and intervention. Appl. Psychol. Int. Rev. **54**(2), 199–231 (2005)
12. Tabuenca, B., Kalz, M., Drachsler, H., Specht, M.: Time will tell: the role of mobile learning analytics in self-regulated learning. Comput. Educ. **89**, 53–74 (2015)
13. Sha, L., Looi, C., Chen, W., Zhang, B.H.: Understanding mobile learning from the perspective of self regulated learning. J. Comput. Assist. Learn. **4**(28), 366–378 (2012)
14. Boekaerts, M.: Self-regulated learning: where we are today. Int. J. Educ. Res. **31**(6), 445–457 (1999)
15. Pintrich, P.R.: The role of goal orientation in self-regulated learning. In: Boekaerts, M., Pintrich, P.R., Zeidner, M. (eds.) Handbook of Self-Regulation, pp. 451–502. Academic Press, San Diego (2000)
16. Weinstein, C.E., Husman, J., Dierking, D.R.: Self-regulation interventions with a focus on learning strategies. In: Boekaerts, M., Pintrich, P.R., Zeidner, M. (eds.) Handbook of Self-Regulation, pp. 727–747. Academic Press, San Diego (2000)
17. Zimmerman, B.J.: Investigating self-regulation and motivation: historical background, methodological developments, and future prospects. Am. J. Int. Res. **45**, 166–183 (2008)
18. Kizilcec, R.F., Pérez-Sanagustín, M., Maldonado, J.J.: Self-regulated learning strategies predict learner behavior and goal attainment in massive open online courses. Comput. Educ. **104**, 18–33 (2017)
19. Panadero, E.: A review of self-regulated learning: six models and four directions for research. Front. Psychol. **8**, 422 (2017)
20. Panadero, E., Broadbent, J.: Developing evaluative judgement: a self-regulated learning perspective. In: Boud, D., Ajjawi, R., Dawson, P., Tai, J. (eds.) Developing Evaluative Judgement: Assessment for Knowing and Producing Quality Work. Routledge, Abingdon (2018)
21. Vohs, K.D., Baumeister, R.F., Schmeichel, B.J., Twenge, J.M., Nelson, N.M., Tice, D.M.: Making choices impairs subsequent self-control: a limited-resource account of decision making, self-regulation, and active initiative. Motiv. Sci. **1**(S), 19–42 (2014)
22. Wijekumar, K., Ferguson, L., Wagoner, D.: Problem with assessment validity and reliability in web-based distance learning environments and solutions. J. Educ. Multimed. Hypermedia **15**, 199–215 (2006)
23. Barnard, L., Lan, W.Y., To, Y.M., Paton, V.O., Lai, S.L.: Measuring self-regulation in online and blended learning environments. Internet High. Educ. **12**(1), 1–6 (2009)

24. Loyens, S.M.M., Magda, J., Rikers, R.M.J.P.: Self-directed learning in problem-based learning and its relationships with self-regulated learning. Educ. Psychol. Rev. **20**(4), 411–427 (2008)
25. Broadbent, J., Poon, W.L.: Self-regulated learning strategies & academic achievement in online higher education learning environments: a systematic review. Internet High. Educ. **27**, 1–13 (2015)
26. Chang, C.C., Liang, C., Shu, K.M., Tseng, K.H., Lin, C.Y.: Does using e-portfolios for reflective writing enhance high school students' self-regulated learning? Technol. Pedagogy Educ. **25**(3), 317–366 (2016)
27. Schunk, D.H.: Self-regulated learning: the educational legacy of Paul R. Pintrich. Educ. Psychol. **40**(2), 85–92 (2005)
28. Puzziferro, M.: Online technologies self-efficacy and self-regulated learning as predictors of final grade and satisfaction in college-level online courses. Am. J. Distance Educ. **22**(2), 72–89 (2008)
29. Reason, P., Bradbury, H.: Handbook of Action Research. Sage, London (2001)
30. Eden, C., Huxham, C.: Action research for the study of organisation. In: Clegg, S., Hardy, C., Nord, W. (eds.) The Handbook of Organisation Studies, pp. 526–542. Sage Publications, London (1996)
31. Goldstein, B., Ick, M., Ratang, W., Hutajulu, H., Blesia, J.U.: Using the action research process to design entrepreneurship education at Cenderawasih University. En: Procedia Social and Behavioral Sciences, vol. 228, Valencia, Lexington (2016)
32. Barnard-Brak, L., Lan, W.Y., Paton, V.O.: Profiles in self-regulated learning in the online learning environment. Int. Rev. Res. Open Distance Learn. **11**(1), 61–80 (2010)
33. Broadbent, J.: Comparing online and blended learner's self-regulated learning strategies and academic performance. Internet High. Educ. **33**, 24–32 (2017)
34. Basaglia, S., Caporarello, L., Magni, M., Pennarola, F.: Individual adoption of convergent mobile phone in Italy. RMS **3**, 11–18 (2009)
35. Al-Busaidi, K.A.: Learners' perspective on critical factors to LMS success in blended learning: an empirical investigation. Commun. Assoc. Inf. Syst. **30**(2), 11–34 (2012)
36. Manolis, C., Burns, D.J., Assudani, R., Chinta, R.: Assessing experiential learning styles: a methodological reconstruction and validation of the Kolb Learning Style inventory. Learn. Individ. Differ. **23**, 44–52 (2013)

Recommending Tasks in Online Judges

Giorgio Audrito[1,3], Tania Di Mascio[2], Paolo Fantozzi[4], Luigi Laura[3,4](✉), Gemma Martini[3], Umberto Nanni[4], and Marco Temperini[4]

[1] University of Torino, Turin, Italy
giorgio.audrito@unito.it
[2] University of L'Aquila, L'Aquila, Italy
tania.dimascio@univaq.it
[3] Italian Association for Informatics and Automatic Calculus (AICA), Milan, Italy
martini.gemma3@gmail.com
[4] Sapienza University of Rome, Rome, Italy
{fantozzi,laura,nanni,marte}@diag.uniroma1.it

Abstract. Online Judges are e-learning tools used to improve the programming skills, typically for programming contests such as International Olympiads in Informatics and ACM International Collegiate Programming Contest.

In this context, due to the nowadays broad list of programming tasks available in Online Judges, it is crucial to help the learner by recommending a challenging but not unsolvable task. So far, in the literature, few authors focused on Recommender Systems (RSs) for Online Judges; in this paper we discuss some peculiarities of this problem, that prevent the use of standard RSs, and address a first building brick: the assessment of (relative) tasks hardness.

We also present the results of a preliminary experimental evaluation of our approach, that proved to be effective against the available dataset, consisting in all the submissions made in the Italian National Online Judge, used to train students for the Italian Olympiads in Informatics.

Keywords: Recommender systems · Programming contests · e-learning

1 Introduction

In recent years we have witnessed the diffusion of Programming Contests (PCs), i.e. competitions in which participants are faced a set of tasks that require writing computer programs. The importance and the effectiveness of PCs in the process of learning computer programming and, more generally, computer science has been broadly emphasized [3–5,8,12,15].

Training for PCs relies heavily on Online Judges (OJs), also called Programming Online Judges, that are web based e-learning tools where a learner can submit solutions to a programming task. The learner chooses a task and reads its statement; then, online or offline, he writes a solution that is submitted to the

R. Gennari et al. (Eds.): MIS4TEL 2019, AISC 1007, pp. 129–136, 2020.
https://doi.org/10.1007/978-3-030-23990-9_16

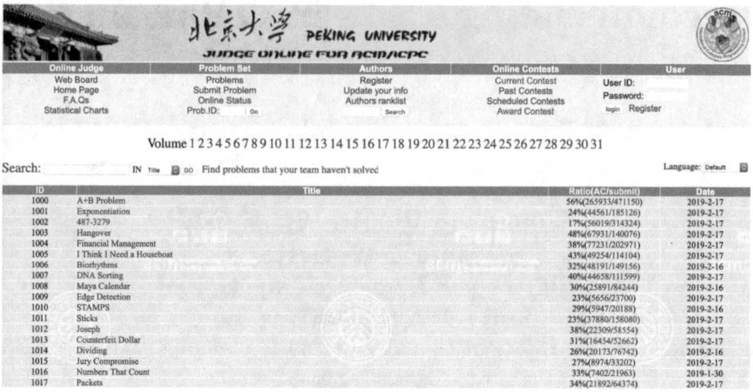

Fig. 1. The list of available problems in the Peking University OJ.

OJ, that verifies the correctness, usually by testing it against a certain number of test cases, and the efficiency, by checking that the running time and/or the memory usage is under some limit.

However, as observed in [17], the large number of tasks available to users is a typical example of *information overloading scenario*: an unexperienced user has to choose from thousands programming tasks, many of which are probably too difficult for him. Just to provide some examples, University of Valladolid Online Judge has more than 200k users and 2k tasks, whilst SPOJ accounts approximately 600k users and 6k (public) tasks. In Fig. 1 is shown the typical interface with the list of tasks in an OJ platform, whilst in Fig. 2 is shown an example of a programming task.

With such numbers, a Recommender System (RS) for the users is definitely needed, to help them finding the next task. Traditional RS approaches can be very broadly divided into two categories: Content Based ones, in which the recommendations derive from features of the items to be suggested, and Collaborative Filtering approaches, in which the suggestion is based on the items chosen by users *similar to the current one.*

There are, however, some peculiarities of Online Judges that prevent the use of a general Recommender System:

- the user slowly improves his abilities, one task after the other, so the general concept of user *preferences* does not apply: recommending a movie or a book differs significantly from recommending a task; a user will probably still like a movie after one year, whilst he might find a task too easy after the same amount of time.
- Users with *similar* skills, i.e. users to whom we might want to suggest the same set of tasks, might behave very differently in OJs, thus preventing us from considering them *similar.* For example, one might solve all the tasks involving a given skill, while the other might just solve one task, related to that skill, and then move on to tasks involving different skills.

Note that the above issues are typical of RSs in e-learning tools, thus successful approaches in this field might be extended to more general cases of TEL systems.

In this paper we propose an algorithmic approach to a building block of a task recommender system: given a list of tasks solved by the users, we want to estimate relative hardness of the tasks, i.e. finding a ranking of the tasks from the easiest to the hardest.

Notice that the number of users who solved a given task, which could be seen as a proxy for the hardness of the task, is not a good indicator: popular hard tasks might have more users that solved them compared to easy, unpopular, tasks.

Our approach is based on the construction of a graph, where the nodes are the tasks and the (weighted) directed edges represent the number of users that solved one task before the other. We tested the effectiveness of our approach on the data from the OJ used by the secondary school students training for the Italian Olympiads in Informatics (Olimpiadi Italiane di Informatica - OII) [10], and the preliminary experimental results confirm the effectiveness of our approach.

Railway Schedule (`paths`)

The railway network in the *Pordenone* county consists of N train stations connected by $N-1$ tracks (X_i, Y_i) so that from every station is possible to reach any other station: in other words, the tracks form a *tree*.
This choice makes the transportation system extremely inefficient: trains going in opposite directions cannot cross each other on a single track, so they need to perform lengthy and complex manoeuvres to pass each other. The new administration founded its campaign trail on changing this situation once and for all... and now it's time to keep promises!
Edoardo, the local leading expert in logistics, already has a mind-blowing idea for fixing the situation: making each track one-way, so that no crossings will ever occur! Of course, the tricky part is choosing the orientations so that the service remains acceptable for the majority of the population. After inspecting the traffic patterns, Edoardo discovered that most people travel between one of M pairs (A_i, B_i) of stations. Thus, an orientation of the tracks will be considered *acceptable* by the population only if for each such pair, either a path from A_i to B_i or a path from B_i to A_i should exist. However, many acceptable orientations exist and Edoardo cannot choose among them, otherwise his system would be deemed as unfair: the only solution is to use *all* of them in a periodic schedule of daily track orientations. Help Edoardo design such a schedule by counting how many acceptable orientations exist! Since this number may be large, report it modulo 1 000 000 007.

Fig. 2. An example of a problem from a programming contest; this task is taken from the final contest of the 2019 edition of the Italian Team Olympiads in Informatics (OIS) [2].

This paper is organized as follows: the next section provides the necessary background related to programming contests, online judges, and recommender systems, whilst our approach is detailed in Sect. 3. In Sect. 4 we describe our experimental findings and concluding remarks are addressed in Sect. 5.

2 Background

In this section we provide the reader the necessary background concerning programming contests, online judges, and recommender systems.

2.1 Programming Contests

A programming contest is a competition in which contestants are faced with a set of programming tasks, also called problems, to be solved in a limited amount of time and/or with a limited amount of memory usage.

A single task can be broken into different subtasks of increasing complexity: basic techniques might be enough to solve, within the given time and/or space limits, some of the subtasks whilst the most difficult ones might require very specific algorithmic techniques and data structures.

We mention some popular programming contests:

- The International Olympiads in Informatics (IOI), that are an annual programming competition for secondary school students patronized by UNESCO. http://www.ioinformatics.org/
- The ACM International Collegiate Programming Contest (ICPC) is a multi-tier, team-based, programming competition operating under the auspices of ACM. https://icpc.baylor.edu/
- The very recent International Olympiads in Informatics in Team (IOIT), that started in 2017, that are a team competition, like ACM ICPC, differently from IOI (individual competition). Currently there are only four nations involved: Italy, Romania, Russia, and Sweden. https://ioi.team/
- Google Code Jam, that is based on multiple online rounds that concludes in the World Finals. https://code.google.com/codejam/.
- Facebook Hacker Cup, that is *an annual worldwide programming competition where hackers compete against each other for fame, fortune, glory and a shot at the coveted Hacker Cup.* https://www.facebook.com/hackercup/

2.2 Online Judges

The Online Judges are, usually, web based platforms that provide a large number of programming tasks to be solved. There are several popular OJ platform, we cite the already mentioned University of Valladolid Online Judge https://uva.onlinejudge.org, Sphere Online Judge (SPOJ) https://www.spoj.com/, CodeChef https://www.codechef.com/, and Peking University Online Judge http://poj.org.

In the literature, the first reference to Online Judge dates to the paper of Kurnia, Lim, and Cheang [13]. A brief survey on OJs can be found in [17], whilst more extended surveys on tools and techniques for automatic evaluation of solutions submitted to OJs can be found in [1,6].

2.3 Recommender Systems in OJs

As already observed in the introduction, despite the large amount of literature devoted to RS, the peculiarities of recommendation in OJs, where the relation user-item is way more complex than the typical RS cases, prevent from using standard techniques and forces the development of ad-hoc methods.

However, so far few research focused in the recommendation of tasks in OJs: in [14] the authors use the traditional collaborative filtering method with a new similarity measure adapted to the case, whilst in [17] is presented an approach based on fuzzy logic, refining a previous approach [16]. In [7], Caro and Jimenez tackled the problem by considering user-based and similarity-based approaches. An alternative approach is detailed in [11], where is defined a framework that can allow recommendations and that can foster motivation in students by means of a lightweight, badge-based, gamified approach.

Our approach differs from the ones cited above because we aim at solving a subproblem: can we derive the ranking of the tasks, ordered by their hardness?

3 Ranking Tasks in Online Judges

In this section we details our approach. Our goal is to provide a rank of the tasks, based only on the submissions made by the users. Thus, our input data contain all the submissions to an OJ platform. Our approach is based on the following assumptions:

- a task is solved by a user if and only if he has obtained the maximum possible score on it
- having two tasks t_1 and t_2 where t_1 is harder than t_2, then each user, most of the times, will solve them in order of ascending difficulty (so first t_2 and the t_1)
- it is possible to estimate the difficulty of a task, just using the users' submissions, without any knowledge about the users

The submissions are sorted by user and timestamp, to have the ordered sequence of the tasks for each user. Since we assume that each user solves the tasks in ascending order of difficulty, then we can consider this sequence as monotonically increasing in terms of difficulty.

So, considering d_i as the measure of the difficulty of t_i, we assume that in each sequence $d_i \leq d_j \iff i < j$. Now we build a weighted directed graph with **tasks as nodes** and where each edge in the graph (t_i, t_j, w) means that $d_i < d_j$ in w different sequences.

Since that it is possible that if a task could be solved by some users just after it has been uploaded on the platform, even if it is not harder than the last one solved by the user, then we try to reduce the error induced by this, avoiding counting the sequences where the timestamp of t_i is prior than the uploading of t_j on the platform in the weight of the edge (t_i, t_j).

Since that a sequence of length n will create $O(n^2)$ edges, we expect to have a much dense graph. Moreover, an old task (uploaded at the beginning of the utilisation of the platform) will likely have more out edges than the newer tasks. To decrease the bias given by the age of the tasks, we create N random walks on the graph with random length, chosen in rl_m and rl_M.

Each random walk starts from a random node, and then, in each iteration a random out edge e_i of the node n is chosen such that the probability p of choosing e_i is equal to

$$p(e_i) = \frac{w_i}{\sum\limits_{j \mid e_j \in out(n)} w_j}$$

At this point we have N sequences such that:

$$\forall\, t_i, t_{i+1} \in S \implies (t_i, t_{i+1}) \in G$$

where S is the set of N random walks' paths, G is the original graph built from submissions. So we build a subgraph $G' = (V', E')$ of $G = (V, E)$ such that

$$\forall (s, d, w') \in E' \implies (s, d, w) \in E,\ w' = |\{p \mid (t_i, t_{i+1}) \in p,\ p \in S,\ t_i = s,\ t_{i+1} = d\}|$$

The resulting graph should have the same order of magnitude for all the edges. This means that the bias given by the age of the task is reduced drastically. Then, the edges are once again filtered out leaving only one direction. In practice we will maintain only the edge with the maximum weight between $s \rightarrow d$ and $d \rightarrow s$.

This final graph is used to define an order between nodes, using different metrics. For example a score for a node n:

$$m(n) = \frac{\sum\limits_{i \mid e_i \in in(n)} w_i}{\sum\limits_{j \mid e_j \in out(n)} w_j + \sum\limits_{i \mid e_i \in in(n)} w_i}$$

4 Experimental Evaluation

In order to evaluate the effectiveness of our approach, we considered a dataset with the data from the OJ used by the secondary school students training for the Italian Olympiads in Informatics (Olimpiadi Italiane di Informatica - OII) [9,10].

In particular, this dataset had:

- 321430 submissions
- 366 tasks
- 3928 users

We considered only the submissions that solved the tasks, so we reduced to 68859 submissions, where the distributions of the users with respect to the tasks are shown in Fig. 3. We computed 100000 random walks, that is a 10-factor over the number of edges of the starting graph. The resulting graph contains 366 nodes (i.e., one for each task in the OJ platform) and 121803 edges.

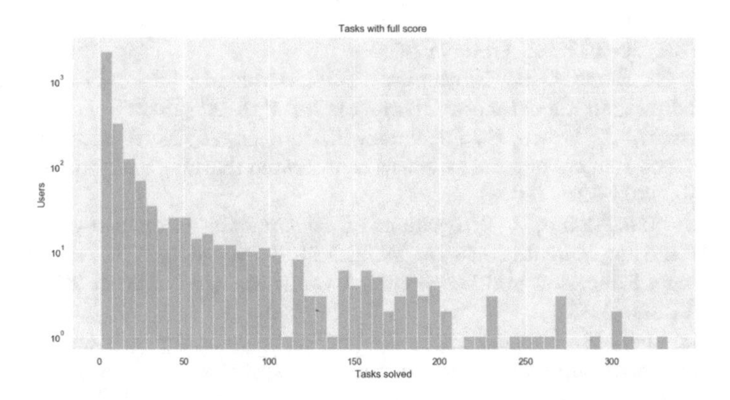

Fig. 3. Distribution of number of solved tasks per users

Judging the hardness of a task is, by definition, a subjective problem. In order to assess our results we only considered the top 25 tasks, and divided them into five buckets of five tasks each. We asked some three experts (i.e. the tutors that maintain the platform) to evaluate our results by sort the five buckets in the order of the hardness of the tasks included. Two experts sorted the buckets in the same order obtained by the algorithm, whilst the third one swapped the third and fourth bucket. We plan to test our approach on data from other OJs, but the validation of the results is a complex issue by itself.

5 Conclusions

In this paper we proposed a graph based approach to estimate the relative hardness of tasks in OJs. This is a basic building block of a recommending system to suggest the next task to be solved by a user.

We also performed an experimental evaluation of our approach against the data from the OJ used by the secondary school students training for the Italian Olympiads in Informatics (Olimpiadi Italiane di Informatica - OII) [9,10].

Our preliminary results seem promising, and we plan to carry on our investigations by testing it with different data; furthermore, as mentioned in the previous section, the problem of the evaluation of the results is a complex task by itself, and we plan to try alternative approaches also in this directions, including a comparison with traditional recommendation methods (including random orders).

References

1. Ala-Mutka, K.M.: A survey of automated assessment approaches for programming assignments. Comput. Sci. Educ. **15**(2), 83–102 (2005)
2. Amaroli, N., Audrito, G., Laura, L.: Fostering informatics education through teams olympiad. Olympiads Inf. **12**, 133–146 (2018)
3. Astrachan, O.: Non-competitive programming contest problems as the basis for just-in-time teaching. In: 2004 34th Annual Frontiers in Education, FIE 2004, vol. 1, pp. T3H/20–T3H/24, October 2004
4. Audrito, G., Demo, G.B., Giovannetti, E.: The role of contests in changing informatics education: a local view. Olympiads Inf. **6**, 3–20 (2012)
5. Blumenstein, M., Green, S., Fogelman, S., Nguyen, A., Muthukkumarasamy, V.: Performance analysis of game: a generic automated marking environment. Comput. Educ. **50**, 1203–1216 (2008)
6. Caiza, J., Del Alamo, J.: Programming assignments automatic grading: review of tools and implementations. In: INTED 2013 Proceedings of 7th International Technology, Education and Development Conference, 4–5 March 2013, pp. 5691–5700. IATED (2013)
7. Caro-Martinez, M., Jimenez-Diaz, G.: Similar users or similar items? Comparing similarity-based approaches for recommender systems in online judges. In: Aha, D.W., Lieber, J. (eds.) Case-Based Reasoning Research and Development, vol. 10339, pp. 92–107. Springer, Cham (2017)
8. Dagienė, V.: Sustaining informatics education by contests. In: International Conference on Informatics in Secondary Schools-Evolution and Perspectives, pp. 1–12. Springer (2010)
9. Di Luigi, W, Fantozzi, P., Laura, L., Martini, G., Morassutto, E., Ostuni, D., Piccardo, G., Versari, L.: Learning analytics in competitive programming training systems. In: 2018 22nd International Conference Information Visualisation (IV), pp. 321–325, July 2018
10. Di Luigi, W., Farina, G., Laura, L., Nanni, U., Temperini, M., Versari, L.: Oii-web: an interactive online programming contest training system. Olympiads Inf. **10**, 195–205 (2016)
11. Di Mascio, T., Laura, L., Temperini, M.: A framework for personalized competitive programming training. In: 2018 17th International Conference on Information Technology Based Higher Education and Training (ITHET), pp. 1–8, April 2018
12. Garcia-Mateos , G., Fernandez-Aleman, J.L.: Make learning fun with programming contests. In: Transactions on Edutainment II, pp. 246–257. Springer (2009)
13. Kurnia, A., Lim, A., Cheang, B.: Online judge. Comput. Educ. **36**(4), 299–315 (2001)
14. Toledo, R.Y., Mota, Y.C.: An e-learning collaborative filtering approach to suggest problems to solve in programming online judges. Int. J. Distance Educ. Technol. **12**(2), 51–65 (2014)
15. Wang, T., Su, P., Ma, X., Wang, Y., Wang, K.: Ability-training-oriented automated assessment in introductory programming course. Comput. Educ. **56**(1), 220–226 (2011)
16. Yera, R., Martínez, L.: A recommendation approach for programming online judges supported by data preprocessing techniques. Appl. Intell. **47**(2), 277–290 (2017)
17. Yera Toledo, R., Caballero Mota, Y., Martínez, L.: A recommender system for programming online judges using fuzzy information modeling. Informatics **5**(2), 17 (2018)

A Board Game and a Workshop for Co-creating Smart Nature Ecosystems

Rosella Gennari[1], Alessandra Melonio[1], Maristella Matera[2], and Eftychia Roumelioti[1(✉)]

[1] Free University of Bozen-Bolzano, Bolzano, Italy
gennari@inf.unibz.it, alessandra.melonio@unibz.it,
eftychia.roumelioti@stud-inf.unibz.it
[2] Milan Polytechnics, Milan, Italy
maristella.matera@polimi.it

Abstract. Younger generations from urban areas spend an increasing amount of time indoors with technology, e.g., with mobiles. GEKI is an exploratory project that investigates how to co-create with younger generations smart nature ecosystems and get them to spend time outdoors. This paper presents the design of a novel board game and a workshop with it for co-creating such an ecosystem with children.

Keywords: Nature · Children · Workshop · Smart ecosystem · Board game

1 Introduction

Experimental research shows that involving children in outdoor activities in nature is important for their development, and leads to many proven benefits. Firstly, it leads to pro-environmental behaviours, independently of whether children participate in wild environments, or "domesticated" natural environments like parks (e.g., [1,16]). It brings resilience to stress and adversity [4] and improvements in mood of teens [11]. Time outdoors also positively affects physical well being [3]. On the contrary, lack of time spent outdoors in nature can result in mental and physical health issues, e.g., children with Attention Deficit Hyperactivity Disorder (ADHD) who play regularly in green play areas show milder symptoms than those who play in built outdoor and indoor settings [15].

Unfortunately, the quality and amount of time children from urbanised areas spend in natural outdoor environments are dramatically changing. These children tend to spend much more time indoors than outdoors—within schools, childcare centres, gym facilities and vehicles—to the point that Louv refers to them as children with a "nature deficit" [12]. The increased usage of technology for indoor activities, for watching TV, surfing the web, playing video games, is often blamed as one of the main causes of children's living indoors. However,

Supported by the GeKI project.

R. Gennari et al. (Eds.): MIS4TEL 2019, AISC 1007, pp. 137–145, 2020.
https://doi.org/10.1007/978-3-030-23990-9_17

causes are complex, family and society-driven, as indicated in a recent survey of the National Wildlife Federation [5]. Technology, concludes the survey, can be beneficial but it has to be differently used in nature than nowadays. The Get-Out-Kids-and-Interact (GeKI) project considers such concerns and sustains the design of novel smart nature ecosystems with and for children.

GeKI has started investigating how to engage children in the design of novel smart nature ecosystems for them. To this aim, it has created a board game, the Nature Board Game, for children. By playing the game, they learn how to co-create interactive objects for such ecosystems. The paper starts presenting related work in the area of interaction design and children. Then it explains the conception and latest evolution of the game, as used in a workshop with children. Results of the workshop are used to reflect over the idea of a board game for engaging children in the design of smart nature ecosystems for them.

2 Related Work

Research in Human Computer Interaction (HCI) is considering how to design technology that brings children outdoors and adds instead of subtracting from the experience, e.g., see the CHI 2018 workshop [9]. The shared view is that technology should be differently designed for outdoors activities in nature, so as to make children reconsider spending time outdoors. Anggarendra and Brereton consider this to be a prominent research direction: in their conclusions to their HCI literature review [2], they sustain that HCI research for outdoors natural environments should not only be concerned with sustainability issues, but it could also be concerned with engaging people to spend time outdoors per se.

Participatory design or co-design more in general strive to enhance environments where people live by engaging them in designing or re-designing the environments themselves [14]. Although there are several participatory or co-design workshops with cards or other generative toolkits for designing smart or IoT objects with children, e.g., for enhancing cities [13] or for creating socio-emotional bonds [7,8], there are fewer workshops for co-designing with children smart nature ecosystems. An exception is by Smart Toy LLC, which partnered with the National Wildlife Federation in USA to create a mobile gamified app and a companion smart toy that encourage USA children to connect with nature [10]. The design of their solution started with a qualitative study and a participatory workshop, which asked children to imagine their toys "to help others like them to connect with nature". Results of the workshop were paper-based prototypes. The research of GeKI, reported in this paper, moves from similar ideas but goes one step further. It aims to co-create smart nature ecosystems with children by playing with them. The first step in the research is the creation of a board game for co-creating with children, and its usage in workshops with children. Its design and usage are unravelled in the remainder of the paper.

Fig. 1. Deck of Nature Cards, consisting of: Environment Cards (top two rows); Input Cards (middle two rows); Output Cards (bottom row)

3 Board Game

The *Nature Board Game* is a collaborative board game that takes place in the central park of a city, which is immersed in nature. Its goal and how to win it, roles of players, material and rules are described next.

3.1 Goal and Roles

The game is designed for 2–4 players, who are children older than 8, and a person with expertise in interaction design. This person plays the role of *Mayor*. The Mayor wants to organise a festival for children, and needs help to make the park "smart", by enhancing it with interactivity. Players act as *designers* of the new smart nature ecosystem. The game terminates when the goal is reached: each player has created (at least) an interactive object for the park.

3.2 Material

The game per se consists of a Game Board (see Fig. 2), 4 Mission Cards, 1 Mission Board, 4 Tokens, a deck of Nature cards, 1 Table with all the available cards (see Fig. 1), Coins, 1 Dice, 4 Note Sheets and 1 one-minute Hourglass. The deck of Nature Cards consists of: 18 Environment Cards concerning Nature elements or outdoors elements of parks that children can make interactive, e.g., a tree or a swing; 20 Input Cards, concerning triggers for Environment Cards, such as, light sensors; 5 Output Cards, concerning reactions related with Environment Cards, such as lights or sounds.

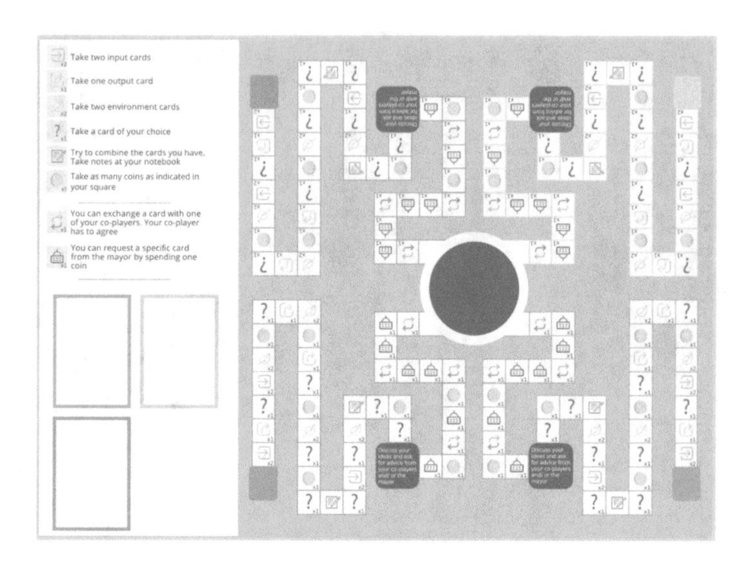

Fig. 2. The board game

Each player takes a note sheet and a token. The *Mayor* keeps a copy of the cards of each category and the coins. The rest of the cards are placed shuffled and faced down on the game board at the corresponding category spot.

3.3 Rules

The *Designers* should carry out one out of the following four missions: (1) add playful and interactive attractions to the park, (2) help visitors explore hidden spots of the park, (3) make sure the visitors respect the park during and after the festival and (4) make sure the park is accessible to everybody. Once a mission is decided with the Mayor, each player places his/her token at a different colorful square (see Fig. 2) and tries to reach the central circle of the board.

Each player, in turn, throws the dice and moves the token the number of spaces indicated by the dice. Depending on the space the token reaches, the player may be entitled to get one or more cards from the piles of cards, get coins, try to combine the cards and take notes, exchange or buy a card. Each time a player's token lands on the notebook space, the *Mayor* turns the hourglass and all the players individually should start thinking of how to combine the cards they already have. They place the cards they do not need on the corresponding place on their note sheet as an indication that they are willing to exchange them. They can also refer to the all-cards table to note down the extra cards they may need. The note sheet and the all-cards table can also be used individually at any time during the game. Among the cards there are blank cards. If a player receives a blank card, he can use it as any other card of his/her choice from the same category.

When a player lands on or passes over the grey square in the middle of his/her path, he/she has to stop there and wait until all the players reach their own grey square. When all the players have reached this point, they have to present in turn their ideas to the rest of the team and the *Mayor*. Each one of the other players should give their opinion and advise the player on how to improve his/her idea. In the end of each turn, the *Mayor* gives the final advice to the player. At the end of this part, each player should have more or less shaped an idea about his/her interactive object and designated on the possible missing cards. After the end of the discussion part, the players keep on moving towards the central circle. When a player lands on the exchange icon, he/she can either exchange or buy a card from another player. When he/she lands on the city-hall icon, he/she can buy a card from the *Mayor*. Each card costs 1 coin. When all the players have reached the center, they should present their final interactive objects with the corresponding cards to the *Mayor* by placing them to the mission board.

4 Workshop

The game was tested and iteratively refined by playing it in 4 main pilot workshops: one with 5 adults, all experts of interaction design; two workshops with 2 11-year-old children and 2 adults with expertise in interaction design; 4 engineering and design students, participating in the Interaction Design Masters course of Milan Polytechnics. Those workshops served to prepare the groundwork for the co-creation workshop described in the remainder of this paper.

4.1 Workshop Description

Exploratory Research Questions. This workshop aimed at exploring whether the board game (1) is understandable for children, (2) is engaging, (3) elicits children's ideas concerning interactive objects for nature outdoors environment.

Fig. 3. Participant players write down their ideas after the completion of the game

Participants and Setting. A workshop was conducted in a house and involved 4 female children, 10–13 years old, one designer, acting as Mayor and moderator, and one instructor experienced in technology-based workshops with children, acting as observer. It lasted circa two hours. All children participated on a voluntary basis and their participation was asked through a consent form.

Protocol. Before starting the game play, children were asked whether they use technological devices in their everyday life and their experience with co-creating interactive solutions. A brief oral presentation was made introducing the definitions of interactive objects, input/sensors, output/actuators and the goal of the workshop. Then the game-play started, divided into two main parts. In the scaffolding part of the game, the input and output cards and the board game were presented and explained to the children. As a first step, the children were asked successively to choose a random card and try to think of what it represents. In case of difficulty in understanding it (e.g., motion input card), the Mayor, acting as moderator, suggested to re-read the description under the title. As a second step, she asked them to take randomly one card of each category and try to think of an interaction scenario. During the game-play part, the game rules were explained and players were given the necessary materials to start. After the completion of the game, the players presented their interactive objects to the Mayor. They were also asked to write down their ideas (see Fig. 3). As a last part, a post-game discussion was held among the players, the Mayor and observer about the game. Players were asked whether they understood and enjoyed the game, as well as to give suggestions for changes.

4.2 Workshop Results

Data related to understandability of the game, engagement with it and ideas emerged from the game play were collected via observations by the moderator (acting as Mayor during the game play) and the observer, as well as via oral questions during the workshop. The moderator and the observer collected their observations independently and then compared their notes, resolving doubts through discussion. Photos were also used to document significant moments. Data are reported below divided per game part: scaffolding; game play.

All players answered that they use technological devices like mobile phones and tablets almost everyday. Two of them (3rd and 4th player) had previous similar experience with robotics workshops. According to the gathered data, nature cards were interpreted without particular problems, especially after reading their description in case of doubts. Among the ideas that came out during the scaffolding part were: a trampoline that lights up when someone jumps on it (*Speed* input card, *Light up* output card), a bird house that vibrates when the bird approaches it (*Distance* input card, *Vibrate* output card). According to the observer and moderator, the scaffolding part helped children break the ice and get them engaged in the game. The players with previous experience seemed more confident during the whole game and helped the others at point, e.g., the former players both made suggestions to the other players.

Table 1. Interactive object ideas.

Players	Interactive objects
1st player (13 years old)	*I have a bike; when it goes fast or when I push a button on it, it plays music*
2nd player (11 years old)	*I am next to a street light that makes a sound when the temperature is high*
3rd player (13 years old)	*The fence has a camera; when it detects people, music is played. The fountain is illuminated with colourful water; when it recognises a colour, a video is projected through water, showing several natural phenomena*
4th player (10 years old)	*When you push the button on the fountain, its lights shine. When the fence has a certain color, it lights up. When the fountain rotates, a light shines*

During the game play, all players showed a good level of enjoyment and overall engagement. Each managed to collaborate and to present at least one idea to the moderator (see Table 1). Mission cards, in particular, were not used in a restrictive way and, in the end, the players seemed to have completely forgotten about them. The final ideas were simple and were mostly based on describing the interaction of few Input/Output Cards with Environment Cards. The suggestions by the children for improving the game were to add more players, more colors, more boxes, a background with trees "to suggest that we are in a real park", and places to illustrate their ideas. Results of the workshop and directions for future work are briefly reflected over in the conclusions.

5 Conclusions and Future Work

The paper reports on the design of the Nature Board Game and its usage in a workshop with children. On the one hand, results of the workshop are limited by its contextual nature, and the fact that it involved only girls. On the other hand, they suggest that children without any experience can understand and engage in the game, and succeed in co-designing interactive objects for smart nature ecosystems. Such objects are however rather simple from the interaction viewpoint.

Based on the observations and players' proposals, on-going work aims at designing a novel version of the game which can further guide children in the construction of smart nature ecosystems. The novel version is divided into different complexity levels. The first level serves to immerse children into the park and explore the usage of cards. Specifically, the first level shows a park with trees to give the feeling of being in nature, as suggested by children. The first level also, with the use of videos, guides children to explore the usage of cards in predefined complex smart objects for nature ecosystems. The other levels instead guide children in their co-construction of their own objects with the Mayor. The final level, in particular, enables children to store their ideas in digital format.

Therefore the board game is turning into a hybrid board game, mixing the physical and the digital, and hence facilitating the sharing of children's novel ideas with other children so as to enable, besides different user classes (like in [6]) collaborative co-design, also their collaborative co-evolution.

References

1. Andrejewski, R., Mowen, A., Kerstetter, D.: An examination of children's outdoor time, nature connection, and environmental stewardship. In: Proceedings of the Northeastern Recreation Research Symposium (2011)
2. Anggarendra, R., Brereton, M.: Engaging children with nature through environmental HCI. In: Proceedings of the 28th Australian Conference on Computer-Human Interaction, OzCHI 2016, pp. 310–315. ACM, New York (2016). https://doi.org/10.1145/3010915.3010981
3. Calogiuri, G.: Natural environments and childhood experiences promoting physical activity, examining the mediational effects of feelings about nature and social networks. Int. J. Environ. Res. Public Health **13**(4), 439 (2016). https://doi.org/10.3390/ijerph13040439
4. Corraliza, J., Collado, S., Bethelmy, L.: Nature as a moderator of stress in urban children. Procedia Soc. Behav. Sci. **38**, 253–263 (2012). https://doi.org/10.1016/j.sbspro.2012.03.347
5. Coyle, K.: Digital technology's role in connecting children and adults to nature and the outdoors (2017). https://www.nwf.org/~/media/PDFs/Kids-and-Nature/NWF_Role-of-Technology-in-Connecting-Kids-to-Nature_6-30_lsh.ashx. Accessed Sept 2018
6. Di Mascio, T., Gennari, R., Melonio, A., Vittorini, P.: The user classes building process in a TEL project. In: Vittorini, P., Gennari, R., Marenzi, I., de la Prieta, F., Rodríguez, J.M.C. (eds.) International Workshop on Evidence-Based Technology Enhanced Learning, pp. 107–114. Springer, Heidelberg (2012)
7. Gennari, R., Melonio, A., Rizvi, M.: The participatory design process of tangibles for children's socio-emotional learning. In: Barbosa, S., Markopoulos, P., Paternò, F., Stumpf, S., Valtolina, S. (eds.) Proceedings of the 6th International Symposium on End-User Development (IS-EUD 2017). LNCS, pp. 167–182. Springer, Cham (2017)
8. Gennari, R., Melonio, A., Rizvi, M., Bonani, A.: Design of IoT tangibles for primary schools: a case study. In: Proceedings of the 12th Biannual Conference on Italian SIGCHI Chapter, CHItaly 2017, Cagliari, Italy, 18–20 September 2017, pp. 26:1–26:6 (2017). https://doi.org/10.1145/3125571.3125591
9. Jones, M.D., Anderson, Z., Häkkilä, J., Cheverst, K., Daiber, F.: HCI outdoors: understanding human-computer interaction in outdoor recreation. In: Extended Abstracts of the 2018 CHI Conference on Human Factors in Computing Systems, CHI EA 2018, pp. W12:1–W12:8. ACM, New York (2018). https://doi.org/10.1145/3170427.3170624
10. Koepfler, J.: Connecting children with nature through smart toy design (2016). https://www.smashingmagazine.com/2016/07/connecting-children-with-nature-through-smart-toy-design/. Accessed Sept 2018

11. Li, D., Deal, B., Zhou, X., Slavenas, M., Sullivan, W.C.: Moving beyond the neighborhood: daily exposure to nature and adolescents'mood. Landsc. Urban Plan. **173**, 33–43 (2018)
12. Louv, R.: Last Child in the Woods. Algonquin Books (2005)
13. Mavroudi, A., Divitini, M., Gianni, F., Mora, S., Kvittem, D.R.: Designing IoT applications in lower secondary schools. In: 2018 IEEE Global Engineering Education Conference (EDUCON), April 2018, pp. 1120–1126 (2018). https://doi.org/10.1109/EDUCON.2018.8363355
14. Sanders, E.B., Stappers, P.J.: Co-creation and the new landscapes of design. CoDesign Int. J. CoCreation Des. Arts **4**(1), 5–18 (2008)
15. Taylor, A., Kuo, M.: Could exposure to everyday green spaces help treat ADHD: evidence from children's play settings. Appl. Psychol. Health Well-Being **3**, 281–303 (2011). https://doi.org/10.1111/j.1758-0854.2011.01052.x
16. Wells, N.M., Lekies, K.S.: Nature and the life course: pathways from childhood nature experiences to adult environmentalism. Child. Youth Environ. **16**(1), 1–24 (2006). http://www.jstor.org/stable/10.7721/chilyoutenvi.16.1.0001

On the Importance of the Design of Virtual Reality Learning Environments

Diego Vergara[1], Manuel Pablo Rubio[2(✉)], Miguel Lorenzo[2], and Sara Rodríguez[2]

[1] Catholic University of Ávila, Ávila, Spain
diego.vergara@ucavila.es
[2] University of Salamanca, Salamanca, Spain
{mprc,mlorenzo,srg}@usal.es

Abstract. Students consider spatial comprehension as one of the major difficulties in engineering studies. A good example of this is the teaching-learning process of ternary phase diagrams (TPDs) where students′ spatial abilities are put to the test. To solve this problem, two virtual learning environments (VLEs) based on virtual reality (VR) are presented in this paper. In essence, they consist of 3D interactive applications designed for interacting in real time with a TPD in different ways: rotating view, exploding view phases, changing point of view, observing hidden zones of the TPD by applying transparency option, cutting the diagram revealing isothermal sections, etc. According to students' opinion shown in this paper, the usefulness of VR in topics that exhibit spatial comprehension difficulties is revealed. Furthermore, comparing the results using each VLE -one developed using technologies from several years ago and the other one using a more updated technology-, the students′ opinion reflects the importance of the VLE design on the motivation that this type of didactic tools awakens in students for being used.

Keywords: Virtual reality · Ternary phase diagrams · Spatial comprehension

1 Introduction

The term technology-enhanced learning (TEL) describes any technology enhancing the teaching-learning process. Many research papers suggest that TEL design is an important aspect to motivate students to learn [1–6]. Motivation in learning is a topic highly analysed in education [7, 8] because of it links with academic performance. Thus, this paper deals with the importance of the design of virtual learning environments (VLEs). To this end, two VLEs based on virtual reality (VR) are considered in the present study. One of them (VLE-1) was developed by using technologies from several years ago [9, 10] and the other one (VLE-2) was recently designed and, hence, a more updated technology was used. Both VLEs deals with the same topic: ternary phase diagrams (TPD), which is part of the contents in diverse engineering degrees (especially in Mechanical Engineering Degree and Materials Engineering Degree). The VLE-1 was applied in class during several years (2013–2017) and the second one was used in class the previous academic course (2017–2018).

R. Gennari et al. (Eds.): MIS4TEL 2019, AISC 1007, pp. 146–152, 2020.
https://doi.org/10.1007/978-3-030-23990-9_18

The aim of this paper is comparing and analysing the students' opinion about both VLEs, thereby revealing the influence of several design aspects on the motivation of using a TEL. Among others, any interesting aspects are: use of up-to-date technologies, level of interactivity, easiness of use, etc. Thus, in this paper is shown how such aspects can affect the students' opinion during the teaching-learning process. Furthermore, the importance of TEL designing is revealed in this paper, since is directly related to students' motivation.

2 Ternary Phase Diagrams

The study of TPD presents serious difficulties of spatial comprehension, since this type of diagrams are in 3D and students must visualize hidden parts into the TPD. A widespread practice to solve this problem is working with isothermal sections for a given temperature or simplify TPD to a binary phase diagrams (BPD) just considering an isoconcentration section [11]. These simplifications are assumed not only in the educational field but also in professional engineering field [12].

Commonly, the traditional methodology to deal with TPD does not include interactive resources in the teaching-learning process. This means a problem for both instructors and students. On one hand, instructors must explain a topic that requires spatial abilities for an adequate understanding. On the other hand, the students face the challenge of understanding a spatial object (TPD) from a bidimensional view (TPDs represented in an 2D image). Thus, VLEs are a really interesting alternative in those cases related where spatial comprehension difficulties are involved, since they allow users to freely interact with virtual objects. This way, there are many examples of using TELs to solve the difficulties in spatial comprehension [13–17].

3 Virtual Reality Learning Environments

Although the two VLEs compared in this paper were designed following the same workflow summarized in the Fig. 1 [18] –similar to those used in previous studies [19–23]–, each of the VLEs has different features. The creation process of both VLEs is based on similar assumptions: (i) 3D modelling of the scene objects, (ii) materials assignment, (iii) lightening and (iv) interactivity programming. However, the design process is different in both the software used and the work technics used. In the design of the VLE-2 the key point was the improvement of two aspects of the VLE-1: the graphical aspect and the interactivity. Thus, the new virtual tool (VLE-2) is more attractive, with a more intuitive use and a better understanding of the key concepts of the TPD analysed. To achieve that goal, different software tools which are currently used in videogames development were used.

For creating the VLE-1 two 3D Windows-based development tools were used: (i) RTre: 3DStudio MAX plugin, and (ii) Quest 3D from the company ACT3D: basic application for generating simple interactive environments. These softwares do not allow to create very realistic environments and they are very limited in lightening and materials libraries. For these reasons, in the VLE-1 only the TPD is shown in an empty environment with a blue background and a white gird representing the ground (Fig. 2a). The lightening and the materials use plain colours without textures, so the level of realism is low. On the other hand, the interaction with the TPD is limited:

(i) the user movements are limited to the TPD surroundings, (ii) the diverse phases only can be separated and coupled (Fig. 2a-left), and (iii) the isothermal sections are simulated with lines.

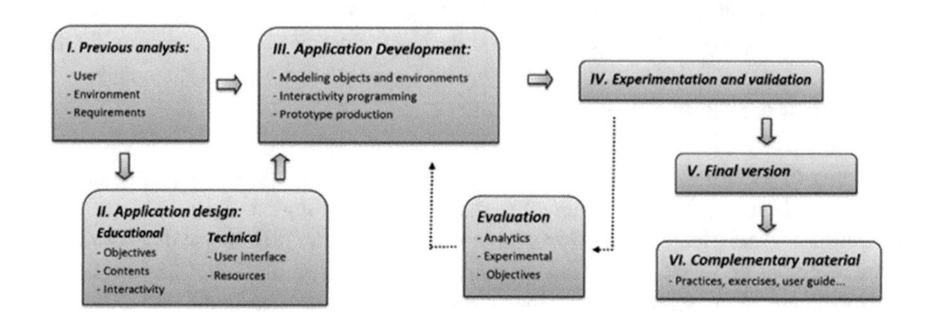

Fig. 1. Workflow in the creation of a VLE [18].

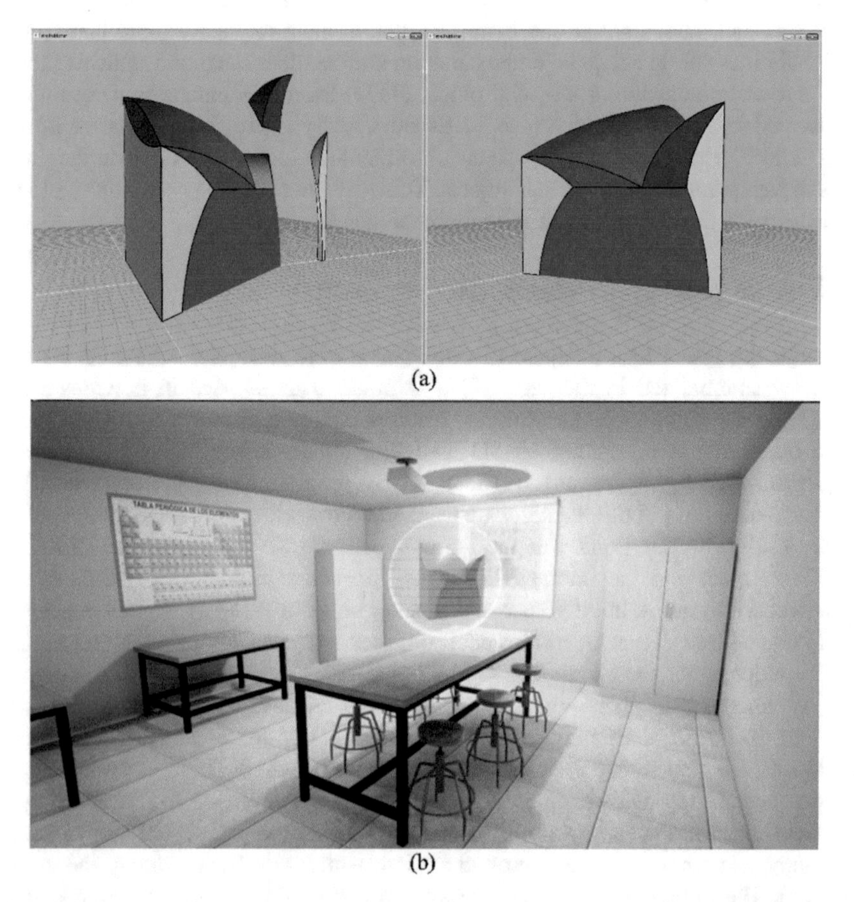

Fig. 2. VLEs based on virtual reality: (a) VLE-1 (old version); (b) VLE-2 (new version).

The VLE-2 has been developed with a more sophisticated software, the well-known Unreal Engine 4. This software is a last generation videogames and interactive applications development engine that, in addition, allows to create immersive and non-immersive virtual reality environments to be used in different platforms (Windows, IOs, Android and videogames console). With Unreal Engine 4, more realistic (more attractive due to the advanced use of lightening and material textures) and more interactive environments can be designed than the software used in the previous VLE-1. The environment designed for the VLE-2 was a 3D lab, similar to the one of any Engineering School. Lightening and materials were obtained by using the method PBR (Physically Based Rendering), which calculates the light interaction with materials by using physical equations. The user mobility within the environment is complete, and he/she can see the TPD and the lab from any point of view. The TPD was simulated as a hologram, floating on a table (Fig. 2b). This contributes to give a dramatic and attractive effect to an object that is just a graphical representation of a data set (imaginary object) [24–26]. The user interaction with the TPD is complete and the phases can be hidden or showed, separated or joint, etc. Furthermore, user can interactively obtain the isothermal sections.

4 Students' Opinion

This section includes the students′ opinions about both VLEs (VLE-1 and VLE-2). The questions included in the survey were directly related to the design of the VLEs (Table 1). VLE-1 was used in four academic courses (2013-2017) and VLE-2 in the last course 2017-2018. The methodology applied in the class was the following: firstly, the theoretical concepts and exercises of BPD were developed in class (approximately 8 h); secondly, a brief explanation of TPD was given through several practical examples showed in a non-interactive way (approximately 1.5 h); finally, students handled the VLE in an interactive and self-learning way (approximately 0.5 h). After that, the students answered the survey question reflected in the Table 1. The mean of the answers obtained for the different academic courses is shown in Fig. 3.

The evolution of the students' opinion is relevant (Fig. 3): (i) the level of motivation to use the VLE-1 decreased from 2013 to 2017; (ii) the level of interactivity follows the same tendency of the motivation; and (iii) the easiness of use maintained the same rating for both VLEs. Authors consider that, in case of VLE-1, the decreasing trend with time of the motivation and interactivity is directly related to the obsolescence of the VLE (which was designed with obsolete technology and, hence, it presents an old appearance). The same reason helps us to justify the increment of students' valuation the last year, just when the updated version of VLE (VLE-2) was used in class. On the other hand, the easiness of use is not affected by the valuation (Fig. 3), since students are used to handle this type of environments (virtual reality learning environments are designed with the same technology as the video-games and, furthermore, all the surveyed students had already used −or they were using at that moment− other VLEs in class [27–33]).

Table 1. Same examples of survey questions.

Number	Question
1	Value from 1 to 10 the "motivation" generating by the LVE
2	Value from 1 to 10 the "interactivity" of the LVE
3	Value from 1 to 10 the "easiness of use" of the LVE

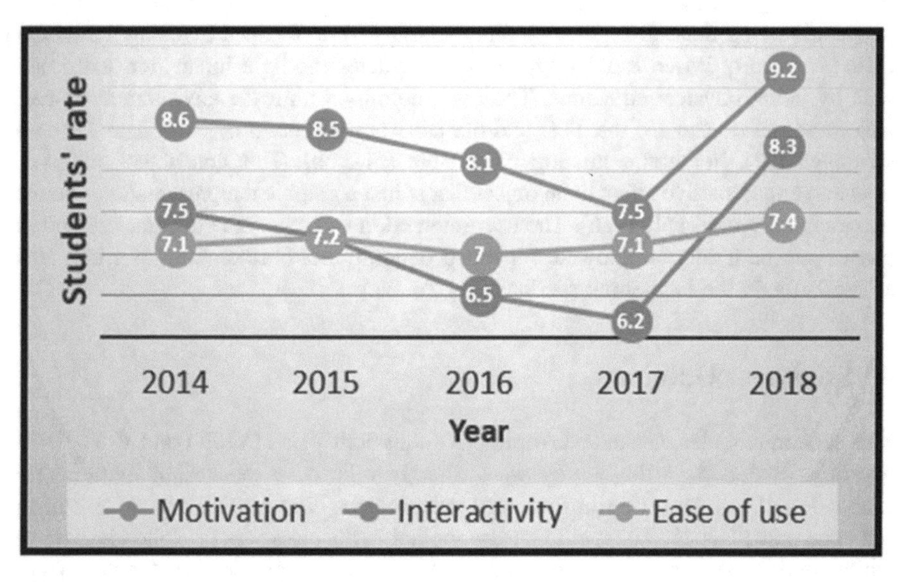

Fig. 3. Students' valuation of the VLEs (VLE-1 from 2014 to 2017 and VLE-2 in 2018)

5 Conclusions

The importance of *using updated technology* for designing a virtual learning environment (VLE) is reflected on the level of motivation that such a didactic resource awakens in students. This paper shows how the level of motivation continually decreases with time when the VLE is becoming obsolete. Thus, the importance of periodically updating a TEL is revealed in this research work, from a teaching-learning point of view. Furthermore, taking into account that student motivation directly influences on the teaching-learning process, teachers should periodically update their TELs so that the didactic tools keep their didactic effectiveness.

As future work, authors will reproduce this study with others VLEs related with Materials Science and Engineering in order to analyse which are the key factors of TEL designing that influence on students' motivation. To this end, to improve the statistical analysis of the present study, a likert scale will be defined and used.

Acknowledgments. This work has been developed as part of "Virtual-Ledgers-Tecnologías DLT/Blockchain y Cripto-IOT sobre organizaciones virtuales de agentes ligeros y su aplicación en la eficiencia en el transporte de última milla", ID SA267P18, project financed by Junta Castilla y León, Consejería de Educación, and FEDER funds.

References

1. Vergara, D., Rubio, M.P., Lorenzo, M.: On the design of virtual reality learning environments in engineering. Multimodal Technol. Interact. **1**, 11 (2017)
2. Chasanidou, D.: Design for motivation: evaluation of a design tool. Multimodal Technol. Interact. **2**, 6 (2018)
3. Vergara, D., Rubio, M.P., Lorenzo, M.: A virtual resource for enhancing the spatial comprehension of crystal lattices. Educ. Sci. **8**, 153 (2018)
4. Chamoso, P., González-Briones, A., Rodríguez, S., Corchado, J.M: Tendencies of technologies and platforms in smart cities: a state-of-the-art review. In: Wireless Communications and Mobile Computing (2018)
5. Palomino, C.G., Nunes, C.S., Silveira, R.A., González, S.R., Nakayama, M.K.: Adaptive agent-based environment model to enable the teacher to create an adaptive class. Advances in Intelligent Systems and Computing, vol. 617 (2017)
6. Griol, D., Molina, J.: Measuring the differences between human-human and human-machine dialogs. ADCAIJ Adv. Distrib. Comput. Artif. Intell. J. **4**, 2 (2015)
7. Conradty, C., Bogner, F.X.: Hypertext or textbook: effects on motivation and gain in knowledge. Educ. Sci. **6**, 29 (2016)
8. LaForce, M., Noble, E., Blackwell, C.: Problem-based learning (PBL) and student interest in stem careers: the roles of motivation and ability beliefs. Educ. Sci. **7**, 92 (2017)
9. Vergara, D., Rubio, M.P., Lorenzo, M.: New virtual application for improving the students′ understanding of ternary phase diagrams. Key Eng. Mater. **572**, 578–581 (2014)
10. Vergara, D., Rubio, M.P., Lorenzo, M.: A virtual environment for enhancing the understanding of ternary phase diagrams. J. Mater. Educ. **37**(3–4), 93–101 (2015)
11. West, D.R.F.: Ternary Phase Diagrams. Chapman & Hall, New York (1982)
12. Chang, Y.-M., Birnie, D.P., Kingery, W.D.: Physical Ceramics: Principles for Ceramic Science and Engineering. Wiley, New York (1996)
13. Rafi, A., Khairul, A., Samad, A., Maizatul, H., Mahadzir, M.: Improving spatial ability using a web-based virtual environment (WbVE). Autom. Constr. **14**, 707–715 (2005)
14. Fonseca, D., Villagrasa, S., Martí, N., Redondo, E., Sánchez, A.: Visualization methods in architecture education using 3D virtual models and augmented reality in mobile and social networks. Procedia Soc. Behav. Sci. **93**, 1337–1343 (2013)
15. Cohen, Ch.A, Hegarty, M.: Visualizing cross sections: training spatial thinking using interactive animations and virtual objects. Learn. Individ. Differ. **33**, 63–71 (2014)
16. Huang, T.Ch., Lin, Ch.Y: From 3D modeling to 3D printing: development of a differentiated spatial ability teaching model. Telemat. Inform. **34**(2), 604–613 (2017)
17. Vergara, D., Rubio, M.P., Lorenzo, M.: On the use of PDF-3D to overcome spatial visualization difficulties linked with ternary phase diagrams. Educ. Sci. **9**(2), 67 (2019)
18. Rubio, M.P., Vergara, D., Rodríguez, S., Extremera, J.: Virtual reality learning environments in materials engineering: Rockwell hardness test. In: Di Mascio, T., et al. (eds.) Methodologies and Intelligent Systems for Technology Enhanced Learning (MIS4TEL 2018). AISC, vol. 804, pp. 106–111. Springer, Cham (2019)

19. Bryson, S.: Approaches to the successful design and implementation of VR applications. In: Virtual Reality Applications, pp. 3–15 (1995)
20. Huang, H.M., Rauch, U., Liaw, S.S.: Investigating learners' attitudes toward virtual reality learning environments: based on a constructivist approach. Comput. Educ. **55**(3), 1171–1182 (2010)
21. García, O., Chamoso, P., Prieto, J., Rodríguez, S., De La Prieta, F.: A serious game to reduce consumption in smart buildings. Commun. Comput. Inf. Sci. **722**, 481–493 (2017)
22. Casado-Vara, R., Prieto-Castrillo, F., Corchado, J.M.: A game theory approach for cooperative control to improve data quality and false data detection in WSN. Int. J. Robust Nonlinear Control **28**(16), 5087–5102 (2018)
23. Casado-Vara, R., Chamoso, P., De la Prieta, F., Prieto, J., Corchado, J.M.: Non-linear adaptive closed-loop control system for improved efficiency in IoT-blockchain management. Inf. Fusion **49**, 227–239 (2019)
24. Erra, U., Malandrino, D., Pepe, L.: A methodological evaluation of natural user interfaces for immersive 3D Graph explorations. J. Vis. Lang. Comput. **44**, 13–27 (2018)
25. Johnston, A.P., Rae, J., Ariotti, N., Bailey, B., Lilja, A., Webb, R., McGhee, J.: Journey to the centre of the cell: virtual reality immersion into scientific data. Traffic **19**(2), 105–110 (2018)
26. Dede, C., Salzman, M.C., Loftin, R.B.: ScienceSpace: virtual realities for learning complex and abstract scientific concepts. In: Proceedings of the IEEE 1996 Virtual Reality Annual International Symposium, pp. 246–252. IEEE (1996)
27. Vergara, D., Rubio, M.P., Lorenzo, M.: Interactive virtual platform for simulating a concrete compression test. Key Eng. Mater. **572**, 582–585 (2014)
28. Vergara, D., Lorenzo, M., Rubio, M.P.: Virtual environments in materials science and engineering: The students' opinion. In: Lim, H. (ed.) Handbook of Research on Recent Developments in Materials Science and Corrosion Engineering Education, 1st edn, pp. 148–165. IGI Global, Hershey (2015)
29. Vergara, D., Rubio, M.P.: The application of didactic virtual tools in the instruction of industrial radiography. J. Mater. Educ. **37**(1–2), 17–26 (2015)
30. Vergara, D., Rubio, M.P., Prieto, F., Lorenzo, M.: Enhancing the teaching/learning of materials mechanical characterization by using virtual reality. J. Mater. Educ. **38**(3–4), 63–74 (2016)
31. Vergara, D., Lorenzo, M., Rubio, M.P.: On the use of virtual environments in engineering education. Int. J. Qual. Assur. Eng. Technol. Educ. **5**(2), 30–41 (2016)
32. Vergara, D., Rubio, M.P., Lorenzo, M.: New approach for the teaching of concrete compression tests in large groups of engineering students. J. Prof. Issues Eng. Educ. Pract. **143**(2), 05016009 (2017)
33. Vergara, D., Rodríguez, M., Rubio, M.P., Ferrer, J., Núñez, F.J., Moralejo, L.: Formación de personal técnico en ensayos no destructivos por ultrasonidos mediante realidad virtual. Dyna **93**(2), 150–154 (2018)

Immersive Virtual Environments: A Comparison of Mixed Reality and Virtual Reality Headsets for ASD Treatment

Tania Di Mascio(✉), Laura Tarantino, Giovanni De Gasperis, and Chiara Pino

Università degli Studi dell'Aquila, 67100 L'Aquila, Italy
{tania.dimascio,laura.tarantino}@univaq.it

Abstract. Since '90s the use of ICT tools has been considered very promising in treatments of people with Autism Spectrum Disorder (ASD) experiencing difficulties in social communication and interaction. Recent literature agrees on potential benefits of Virtual Reality based treatment, particularly in learning processes related to social interaction. The study in this paper describes a usability experience of people with ASD in using Oculus Rift and HoloLens, state-of-the-art Head Mounted Displays (HMDs) allowing users to experience Immersive Virtual Reality and Mixed Reality, respectively. The study focuses on a homogenous target of high-functioning young adults (age 21–23) and aims at evaluating acceptability, usability, and engagement of the two HMDs with respect to such target andproviding general guidelines about applications of the two HMDs.

Keywords: Human-centered computing · Usability testing · Mixed reality · Virtual reality · Laboratory experiments · People with disabilities

1 Introduction

Autism Spectrum Disorders (ASD) are characterized by restricted, repetitive and stereotyped behavior, and core deficits in social communication and interaction [1], which interfere with the process of building relationships and integrating into community [7]. Research has explored a variety of ICT-based approaches to ASD treatment related to social issues as assistive technologies, cognitive rehabilitation tools, and special education tool ([17]). In particular, Virtual Environments (VEs) (i.e., computer-generated representations of environments with realistic appearance) have a great potential for teaching social understanding. This is due to their capability of illustrating scenarios representing situations that may not be feasible in a customary therapeutic setting, promoting role-play, and allowing participants to take a first-person role for skill-learning in a virtual social situation [14]. Immersive Virtual Reality (IVR), Augmented Reality (AR), and Mixed Reality (MR) may be good potential allies for their ability to create Immersive Virtual Environments (IVEs), i.e., synthetic 3D worlds where users may experience a sense of presence tied to the level of visual and interactive fidelity of the IVE with respect to the real world [24]. Existing IVR, AR, and

R. Gennari et al. (Eds.): MIS4TEL 2019, AISC 1007, pp. 153–163, 2020.
https://doi.org/10.1007/978-3-030-23990-9_19

MR devices differ in their ability to match the real world in appearance and functionality, depending on the ability to offer photorealism, to allow users to interact with the IVE elements, or to capture body motion. Engagement (i.e., the ability of physically interact with and control an IVE) is regarded as a key aspect of the sense of presence [2] even if its relationship with immersion is not yet well understood in general literature on IVEs (let alone in studies on IVEs and ASD) [13]. On the other hand, studies about ASD and desktop-based VEs based on physiological markers of engagement (like pupil dilation and blink rate) show that performances of participants improve as engagement increases [9], which suggests that investigation on IVEs is useful and needed in this direction.

Though preliminary studies on VR-based ASD treatment provided positive answers to basic issues related to usability and treatment efficacy (e.g., [8, 10, 21, 23]), the field is far from being mature. Of the 22 pre-post studies reviewed in [17], only one is based on a VE, moreover not immersive. Most of more recent studies on IVEs are based on so called CAVE (cave automatic virtual environment) or semi-CAVE environments (where visual animated images are projected onto the walls and the ceilings of screened space) in which study participants are unable to interact with objects [10, 24]. The same limitation can be found in the study in [8], based on Google Cardboard viewer, a low-cost device based on smartphones creating the illusion of 3D depth and immersion through the stereoscopic effect generated by the biconvex lenses on the VR viewer and the human vision system. In some cases, results are questionable: for example, measures reported in [25] are based on self-rated questionnaires filled out by the children, while studies show that for people with ASD the use of self-report is deprecated, as we discuss later on.

The study reported in this paper focus on the comparison of two state-of-the-art Head Mounted Displays (HMDs), namely Oculus Rift by Oculus VR, subsidy of Facebook, and HoloLens by Microsoft. Both headsets have a high potential for engagement, but differs in the surrounding extent, i.e., the degree to which the physical world is cutout from the view, thus suggesting different roles in ASD treatment. The two HMDs were evaluated in terms of acceptability, usability, and engagement capability with a small group of young people with a diagnosis of ASD, all high-functioning, aged 21–23. Our work differs from other studies (e.g. [14]) in two ways: (1) homogeneity of the users' sample and (2) recourse to objective measures, both particularly significant in the case of ASD people. We notice in particular that the selected age range is currently under-considered in the literature, as testified, e.g., by a recent survey [11]: only 3% of the users collectively evaluated in [11] by the twenty-nine studies examined are of age 20, while none of these studies reported on ASD people older than 20. We hence aim at providing a first contribution in this direction.

This paper is structured as follows: Sect. 2 describes the experiments in terms of evaluated headsets, proposed measure framework, used IVEs scenarios, activities and procedure; Sect. 3 analyzes gathered data and discusses these data in relation to ASD treatment made possible by the considered devices; in Sect. 4 conclusions are drawn.

2 Method

The study was carried out as part of a more general project focused on ICT-based ASD treatments in the areas of communication, social interaction, and autonomy, conducted at TetaLab[1] as a participatory design experience including three computer scientists, four psychologists, one medical doctor, and nine ASD young persons in the age range 15–23. The age range of intended users (older than 13) allows for novel treatments based on state-of-the-art headsets with high levels of immersion and potential for sense of presence, engagement, and expected positive impact on training effectiveness [20]. Since no consolidated results may be found in the ASD literature as to usability/ effectiveness of these devices, we designed a study to investigate their appropriateness.

2.1 Material: Oculus Rift and HoloLens

The *Oculus Rift* headset offers the immersion in a 100% virtual world generated by the processor inside a computer or the VR device while the field of view of the real world is cut out. The *HoloLens* headset adds a layer of AR to the natural field of vision by a translucent visor that allows a vision of the physical world, with virtual elements overlaid on top of it. Oculus Rift and HoloLens are not "devices for all" and have to be used with caution, as indicated by safety warnings [12, 16], with limitations related to age and possible discomfort. HoloLens is not intended for use by children under the age of 13, since an inter-pupillary distance between 51 and 74 is needed to correctly and comfortably view generated holograms, a range accommodating most adults and children older than 13 [12]; Oculus should not be used by children under the age of 13 and children older than 13 should limit the time spent with the device with breaks during use, since prolonged use could negatively impact hand-eye coordination, balance, and multi-tasking ability [16]. Furthermore, even if both devices solve cybersickness problems, it is warned that some people (about 1 in 4000) may have dizziness, seizures, epileptic seizures or blackouts triggered by light ashes or patterns (even if they have no history of seizures or epilepsy), more common in children and people under the age of 20 [12, 14, 23]. Moreover, domain experts report photo sensitivity as characteristics of a meaningful rate of ASD people, even if no consolidated studies or certain prevalence are known. In summary, the use of HMDs should be subject to medical supervision.

2.2 Participants

Six male individuals on the autism spectrum aged 21–23 took part in the study. The participants (all high functioning and attending either high school or University) were recruited within the group of persons involved in TetaLab activities. All participants received a previous diagnosis of ASD (according to DSM V [3]) and none of them had

[1] TetaLab (Technology-Enhanced Treatment for Autism Laboratory) is a laboratory of the University of L'Aquila, based on the cooperation among the Dept. of Information Engineering, Computer Science & Mathematics, the Dept. of Applied Clinical Sciences & Biotechnology, and the University Center for Autism.

a history of epilepsy. Participants' IQ scores ranged from 73 to 98 (M = 85,67, SD = 9,89) according to the Wechsler Adult Intelligence Scale test (WAIS-IV [26]). The participants were required to commit to two sessions on separate days for a duration of approximately 20 min per session plus the time for a short interview to get feedback. The first session, focused on the HoloLens Headset, took place on Oct. 27th in Milan at the Microsoft Lab with the support of two Microsoft people; the second session, focused on the Oculus Rift, was performed on Dec 5th, at TetaLab.

2.3 Measured Factors

The evaluation framework specifically conceived for the study included a number a factors classified in the three macro areas of *acceptability*, *usability*, and *engagement*.

Acceptability was investigated in terms of willingness to use the headset and of a number of factors related to possible unpleasant physiological effects or discomfort (*motion-sickness - MoS*, *double vision - DV*, *digital eye strain - DES*) selected according to literature on IVR and ASD [8, 15], and measured as Boolean values.

Usability was investigated using metrics based on Low (L), Medium (M) and High (H) levels, in terms of: *Autonomy in managing devices* (measuring: support requested to operators during performances - *SO*; autonomy in mounting devices - *MD*); *Comprehension of virtual environment* features (elements – *El*, interactivity – *In*, menu navigation – *Mn*, menu structure – *Ms*); *Interaction ability* in using: game pad – *P*, remote control – *C*, gestures – *G*.

Engagement was investigated using different metrics according to a [1–5] Likert scale, in terms of: *Emotional participation in watching photorealistic images – Ph and non-photorealistic images – NPh*; *Suspension of disbelief* (i.e., extent to which the virtual world is temporarily accepted as reality), *Body participation* (i.e., extent of body movement during the IVE experience), *Exploration of virtual world*; *Input actions*.

2.4 IVEs Scenarios

The Immersive Virtual Environment scenarios administered to participants were selected among demos of the two headsets and applications developed at the DISIM-Univaq ICT Living Lab, based on their appropriateness with respect to the evaluation framework. Demos and applications were preliminarily examined individually by 4 psychologists to prevent uncomfortable or inappropriate experiences for participants.

During the Oculus Rift session students experienced two demos available on the Oculus Store (Introduction to Virtual Reality and the Dreamdeck demos), one application developed at the ICT Living Lab (3D virtual reality model of the Church of Santa Maria Paganica in L'Aquila [6]), and one demo from the Leap Motion Store (the Blocks game); during the HoloLens session students experienced one demo available on the HoloLens Commercial Suite (Michelangelos' David). All scenarios contributed to the evaluation of *acceptability factors*, *autonomy in managing devices*, *comprehension of IVE features*, *suspension of disbelief*, and *body participation*. As to remaining factors, selected scenarios provided different insights according to their characteristics:

- the "Introduction to Virtual Reality" demo allows the user to browse different scenes via remote control, like, e.g., watching the world from the space as if s/he were an astronaut or the far-away lands as if s/he were physically in these lands. These scenarios contributed primarily to evaluating: *emotional participation in photorealistic images, and interaction ability in using remote control* (Fig. 1);
- the Dreamdeck demos offer a mix of photorealistic and not photorealistic scenarios in which the user can, e.g., watch an alien talking with him face to face or forest animals as if s/he were one of them. These scenarios contributed primarily to evaluating: *emotional participation in photorealistic and not photorealistic images*;
- the 3D virtual reality model of the Church of Santa Maria Paganica in L'Aquila, allows the user to explore a historic site. The user can enjoy artistic details like the choir via game pad, approaching them or moving away out of the; the user can also tele-transporting him/her self onto platforms built to observe cupola artistic works from a new point of view. This scenario contributed primarily to the evaluation of: *interaction ability in using the game pad and exploration of the virtual world*;
- in the Blocks game from Leap Motion Store, users, by virtual hands thanks to the Leap motion technology (www.leapmotion.com/), interact with virtual blocks, that can be moved, created them in different geometric forms and levitated. This scenario contributed primarily to the evaluation of *interaction ability via gestures in IVEs*;
- in the Michelangelos' David demo available for the HoloLens Commercial Suite the user can interact with the photorealistic hologram of the statue, miniaturizing it or restoring its original size using his/her own hands, chisel the excess marble, and approach the sculpture to discover the overall artistic details. This scenario contributed primarily to the evaluation of: *photorealistic images, body participation, and interaction ability via gestures in a mixed reality setting*.

2.5 Activities and Procedures

The two evaluation sessions were run as a sequence of customary phases [3, 22]. In a *plenary nurturing phase*, operators introduced themselves, explained the aim of the session, and established a positive atmosphere; during this phase students were told that they could signal discomfort at any time to operators and stop at any time by dismounting the headsets. In the *body phase* (composed of six sub-sessions – one per student) investigators administered activities to students and observed them (for each sub-session three operators, including computer scientists and psychologists, recorded their observations). In a *plenary closing phase*, students had a snack and operators reordered collected material and wrote down their first impressions.

We evaluated the following *activities*: A1: Mounting the HMD; A2: Dismounting the HMD; A3: Browsing menus; A4: Watching virtual environment; A5: Exploring virtual environment; A6: Playing with virtual environment, performed with the same order for each student (Oculus Rift session: A1, A3, A4, A2, A1, A5, A6, A2; HoloLens session: A1, A3, A4, A5, A6, A2). Note that: (1) interaction with HoloLens was only via gestures, while in the case of the Oculus Rift students interacted with IVEs also via game pad and remote control, and (2) in the case of the Oculus Rift participants were asked to dismount and re-mount the headset after the first two

(a) (b)

Fig. 1. Screenshots of the Oculus Rift demos: photorealistic vs non-photorealistic images.

activities in order not to let them idle during the preparation of the following scenario, to ensure a brief rest, and, at the same time, to have an additional test on the autonomy in handling the device.

As *moderating technique*, we used the Concurrent Think Aloud (CTA) technique which allowed us to understand participants' thoughts during their experience by encouraging them to think aloud while they interacted with the IVEs (CTA had no negative side effect on usability evaluation since accuracy and/or of time spent on activities were not to be measured [3, 22]). As to *data collection*, we used the controlled observational method [3]) since both sessions were run in a laboratory setting. To obtain as many accurate data as possible we video-recorded all performances and we interviewed all participants at the end of each performance. Moreover, each operator participating in the evaluation sessions observed students performing experiments and wrote down his/her personal notes. In practice, both direct and indirect observations were used.

From collected materials (notes, and audio/video recording) we extracted quantitative and qualitative data: quantitative data related to *acceptability* were described by Boolean values, quantitative data related to *usability* were described by low, medium and high (L, M, H) scores, and qualitative data for the engagement were described by a [1–5] Likert scale. For *usability factors*, L = v $\in$ [0%…33%], M = v $\in$ [34%…66%], H = v $\in$ [67%…100%] where the meaning of v depends on the particular metrics of the evaluation framework (specifically, *SO*: number (#) of times ASD student asked for support wrt the mean of this measure on Typical Development (TD) people; *MD*: time spent for mounting the HMD wrt the mean of this measure on TD people; *El*: # of virtual elements recognized wrt the total; *In*: # of virtual interactive elements recognized wrt the total: *MN*: # of virtual menu navigation elements recognized wrt the total; *MS*: # of virtual menu structure elements recognized wrt the total; *P*: # of correct actions on the game pad wrt the total requested by the activity; *C*: # of correct actions on the remote control wrt the total requested by the activity; *G*: # correct gestures wrt the total requested by the activity). As to *engagement factors*, the following factors were measured according to a [1–5] Likert scale: degree of *emotional participation* when watching photorealistic (*Ph*)/non-photorealistic (*NPh*) images, degree of *suspension of disbelief*, degree of coherent *body participation* wrt the expected and allowed body movement, degree of *exploration* of the virtual world, degree of participant's input *actions* not required from the activity. As to engagement, we collected both self-reported measures and behavioral observations: self-report measure reflects the perception that a person has about his/her performance of activities [19], while behavioral observational is important for understanding people' actions, roles and behavior [23].

Anyhow it has to be observed that even in the case of TD people results in self-reports may be overestimation/underestimation of actual ability; furthermore, in the specific case of ASD people, self-reports are not recommended for the difficulty that ASD people may have in reflecting and reporting on their own behavior or/and emotions [18]. Indeed, recent literature about empathic ability in individuals with ASD suggests the presence of a dissociation between cognitive and affective empathy in this population: adolescents with ASD feel aroused and involved when others experience emotions as the healthy subjects do. These results suggest that ASD subjects show a difficulty in cognitively identifying and explaining what they feel. For this reason, even if we collected self-reported measures, we do not use them, and we report only data obtained by behavioral observations.

3 Results and Discussion

As to ***acceptability***, all factors received a positive value: all participants were able to mount/dismount the headsets without support and within a time comparable to the time needed by TD people; none of the participants reported/showed negative sensory or physiological experiences. All participants were enthusiastic to participate to both sessions; this is in particular remarkable for the HoloLens session, for which they freely chose to face a somehow burdensome trip to Milan also overcoming physical impairments and/or social phobias. As to ***usability***, we calculated the total success rate per session and the success rate for the main metrics, with the following results:

- *Autonomy in managing devices* – Oculus Rift: 75%, HoloLens: 91%
- *Comprehension of VE features* – Oculus Rift: 60%, HoloLens: 54%
- *Interaction ability* – Oculus Rift: 91%, HoloLens; 75%

As to ***engagement***, in each session and for each student three observers reported a set of Likert values; these data can be mapped onto twelve tables, six per session. A final aggregate summary is given in Table 1.

Achieved results have to be analyzed to single out issues that may affect the success of innovative ASD treatments. Actually, fears on cumbersomeness of HMDs seem to be overcome by our results on acceptability. As to usability factors, the partial success of *comprehension of VE features* is coherent with the need of initial training experienced also by typically developing people attending the lab; the successful scores in *autonomy in managing the device* (particularly for the HoloLens) and *interaction ability* open interesting research lines for innovative interventions. As to engagement, the average scores of *body participation* and *exploration of the virtual world* suggest that practice is needed to feel free to virtually and physically move; the lower score of Oculus wrt HoloLens may be due to the presence of the cable connecting the headset to the computer. *Emotional participation* and *suspension of disbelief* are maybe the most

Table 1. Average of engagement results related to the Oculus Rift (OR) and HoloLens (HL) sessions per engagement metric.

ID	Em. participation		Susp. of disb.	Body part.	Explor.	Action
	Ph	*NPh*				
OR	4,11	4,11	3,18	2,44	3,3	3,58
HL	3,95	–	2,9	3,4	3,25	3,9

interesting results, somewhat deviating from what expected but for this reason providing useful suggestions, respectively for content issues and nature of applications.

Though photorealism is generally suggested to play a critical role in the achievement of engagement (due to the higher fidelity of the IVE wrt the real world [2, 13, 24]), we achieved exactly the same score for *emotional participation* in photorealistic and non-photorealistic scenarios in the Oculus session. This result may be due to the fact that ASD people get easily distracted by irrelevant details of a scene and feel more comfortable in facing simplified reality. Moreover, from individual interviews it clearly emerged that a factor more important than photo-realism for *emotional participation* was the subjective familiarity of the student with IVE objects. This may also the reason why in the HoloLens session the *emotional participation* factor got a lower score.

As to *suspension of disbelief*, generally considered prerequisite for achieving a sense of presence in IVEs [13], it is worth reporting that our study participants underlined a continuous awareness of the distinction between the real and the virtual worlds throughout the duration of the session. For example, when asked whether they were scared, during (possible) unpleasant scenes, they observed that they could not be scared by a fictional world. This difficulty in getting a complete suspension of disbelief is not necessarily negative: it is telling us that for ASD people the IVEs are "safe places" that can be used for training and learning. This, together with the high scores achieved on autonomy and engagement, makes the Oculus Rift a very promising platform for treatment aimed at acquisition/improvement of social skills where performances are a value, provided that personalization of content is possible for a good emotional participation.

Overall, HoloLens achieved lower scores in engagement, particularly in the "suspension of disbelief". This may turn as an advantage if the device is used in applications aimed at the acquisition/improvement of autonomy where the perception of the real world is a value: together with the very high score achieved in "autonomy in using the device" and the fact that HoloLens does not require a physical connection with a computer and can be used as a standalone device, the low degree of suspension of disbelief makes the HoloLens a very promising platform for prosthetic tools able to enrich the real world with objects familiar to the ASD person, making the real world less scaring.

Actually, for ASD individuals the real world is particularly stressful because they have difficulty in understanding social situations/rules. To cope stress and anxiety level, people with ASD avoid altogether the social situation leading to isolation, aloneness, and major dependence by caregivers. A prosthetic application of HoloLens to support ASD individuals during the social situation in real life may function in the same way in which patients with amputations use robotic limbs that respond to a series of contrived operational signals from their body [4]. A HoloLens-based application could represent a social prosthetic treatment supporting ASD individuals during the social situation in real life and allowing them to reach higher levels of independence and functioning.

4 Conclusions

In this paper, we presented a study aimed at comparing two HMDs, Oculus Rift and HoloLens, in terms of *acceptability*, *usability*, and *engagement* with respect to potential application of IVR and MR in technology-enhanced treatment of people with ASD. The results in this study aim at enriching the grounding knowledge needed to conceive technology-enhanced treatment based on state-of-the-art IVR and MR devices. In particular, our results – referred to young high-functioning adults with ASD – suggest that the two evaluated HMDs are accepted and generally usable after short training, which make them promising allies for ASD treatment, in particular for improvements in social skills. The differences, revealed primarily in terms of autonomy and engagement, suggest complementary roles of the two headsets in ASD interventions, with *IVR more appropriate for learning applications*, favored by the higher engagement offered by the high immersion level, and *MR utilized as a prosthetic tool* able to improve the capacity of persons with ASD to cope anxiety, fear and phobias in social situations.

As in other studies in this field, a limitation is given by the small size of the user samples. Actually, as observed also in [11], even if most studies are conducted with small usergroups, this limit is outbalanced by the numbers of studies that collectively contribute to the field. In particular, to the best of our knowledge, our experiment is the first one dealing with young adults in the 20–23 age range.

Finally, we notice that the metrics used in this paper could represent the foundations for an innovative IVE evaluation framework, since differently from customary literature (e.g., [9, 15]), metrics are based on data collected observing ASD people and not on self-reported questionnaires (as observed also in [15]: "...the use of self-reported questionnaires among people with ASD/ID makes the results vulnerable and bias").

Acknowledgment. Authors wish to thank TetaLab people for the fruitful cooperation throughout the study, and Microsoft Italia for their support during the Microsoft sessions in Milan.

References

1. American Psychiatric Association: Diagnostic and Statistical Manual of Mental Disorders: DSM-V. American Psychiatric Publishing, Arlington (2013)
2. Brown, E., Cairns, P.: A grounded investigation of game immersion. In: CHI 2004 Extended Abstracts on HF in CS (CHI EA 2004), pp. 1297–1300. ACM, New York (2004)

3. Cofini, V., Di Giacomo, D., Di Mascio, T., Necozione, S., Vittorini, P.: Evaluation plan of TERENCE: when the user-centred design meets the evidence-based approach. In: Vittorini, P., et al. (eds.) International Workshop on EBUTEL. Advances in Intelligent and Soft Computing, vol 152. Springer, Heidelberg (2012)
4. Cole, J., Crowle, S., Austwick, G., Slater, D.H.: Exploratory findings with VR for phantom limb pain: from stump motion to agency and analgesia. Disabil. Rehabil. **31**(10), 846–854 (2009)
5. Coman, L., Richardson, J.: Relationship between self-report and performance measures of function: a systematic review. J. Aging **25**(3), 253–270 (2006)
6. De Gasperis, G., Mantini, S., Cordisco, A.: The virtual reconstruction project of unavailable monuments: the Church of Santa Maria Paganica in L'Aquila. In: Proceedings of Workshops at the 13th International Conference on Information Theory, pp. 31–33. Springer, Cham (2017)
7. Fletcher-Watson, S., McConnell, F., Manola, E., McConachie, H.: Interventions based on the Theory of Mind cognitive model for autism spectrum disorder (ASD). Cochrane Database Syst. Rev. **21**(3) (2014)
8. Garzotto, F., Gelsomini, M., Occhiuto, D., Matarazzo, V., Messina, N.: Wearable immersive VR for children with disability: a case study. In: Proceedings of the 2017 Conference on Interaction Design and Children (IDC 2017), pp. 478–483. ACM, New York (2017)
9. Lahiri, U., Bekele, E., Dohrmann, E., Warren, Z., Sarkar, N.: A physiologically informed virtual reality based social communication system for individuals with autism. J. Autism Dev. Disord. **45**(4), 919–931 (2015)
10. Maskey, M., Lowry, J., Rodgers, H., McConachie, H., Parr, J.R.: Reducing specific phobia/fear in young people with autism spectrum disorders (ASDs) through a virtual reality environment intervention. PLoS One **9**(7) (2014)
11. Mesa-Gresa, P., Gil-Gómez, H., Lozano-Quilis, J.A., Gil-Gómez, J.A.: Effectiveness of virtual reality for children and adolescents with autism spectrum disorder: an evidence-based systematic review. Sensors **18**, 24–86 (2018)
12. Microsoft: Microsoft HoloLens Health and Safety (2017). https://www.microsoft.com/en-gb/hololens/legal/health-and-safety-information. Accessed 19 Jan 2018
13. Miller, H.L., Bugnariu, N.L.: Level of immersion in VEs impacts the ability to assess and teach social skills in ASD. Cyberpsychol. Behav. Soc. Netw. **19**(4), 246–256 (2016)
14. Mitchell, P., Parsons, S., Leonard, A.: Using virtual environments for teaching social understanding to 6 adolescents with ASD. J. Autism Dev. Disord. **37**(3), 589–600 (2007)
15. Newbutt, N., Sung, C., Kuo, H.J., Leahy, M.J., Lin, C.C., Tong, B.: Brief report: a pilot study of the use of a VR headset in ASD. J. Autism Dev. Disord. **46**(9), 3166–3176 (2006)
16. OculusVR: Oculus Rift Health and Safety (2018). Accessed 19 Jan 2018
17. Parsons, S., Cobb, S.: State-of-the art of virtual reality technologies for children on the autism spectrum. Eur. J. Spec. Needs Educ. **26**(3), 355–366 (2011)
18. Pearl, A.M., Edwards, E.M., Murray, M.J.: Comparison of self-and other-report of symptoms of autism and comorbid psychopathology in adults with autism spectrum disorder. Contemp. Behav. Health Care **2**(1), 1–8 (2016). https://doi.org/10.15761/CBHC.1000120
19. Pino, M.C., Mazza, M.: The use of "Literary Fiction" to promote mentalizing ability. PLoS One **11**(8), e0160254 (2016)
20. Ragan, E.D., Sowndararajan, A., Kopper, R., Bowman, D.A.: The effects of higher levels of immersion on procedure memorization performance and implications for educational virtual environments. Presence (Camb.) **19**(6), 527–543 (2016)
21. Strickland, D.C., Marcus, L.M., Mesibov, G.B., Hogan, K.: Brief report: two case studies using VR as a learning tool for ASD. J. Autism Dev. Disord. **26**(6), 651–659 (1996)
22. Usability.gov: Improving the User Experience (2018). Accessed 19 Jan 2018

23. Tarantino, L., Mazza, M., Valenti, M., De Gasperis, G.: Towards an integrated approach to diagnosis, assessment and treatment in autism spectrum disorders via a gamified TEL system. In: MIS4TEL, pp. 141–149. Springer, Cham (2016)
24. Waller, D., Hunt, E., Knapp, D.: The transfer of spatial knowledge in virtual environment training. Presence (Camb.) **7**(2), 129–143 (1998)
25. Walshe, C., Ewing, G., Griffiths, J.: Using observation as a data collection method to help understand patient and professional roles and actions in palliative care settings. Palliat. Med. **26**(8), 1048–1054 (2012). https://doi.org/10.1177/0269216311432897
26. Wechsler, D.: Wechsler Adult Intelligence Scale, 4th edn. Pearson, San Antonio (2008)

Intelligent Agents System for Adaptive Assessment

Néstor D. Duque-Méndez[1(✉)], Valentina Tabares-Morales[1], and Demetrio A. Ovalle[2]

[1] Universidad Nacional de Colombia, Sede Manizales, Manizales, Colombia
{ndduqueme, vtabaresm}@unal.edu.co
[2] Universidad Nacional de Colombia, Sede Medellín, Medellín, Colombia
dovalle@unal.edu.co

Abstract. Artificial intelligence and especially Multi-Agent systems play a very important role within the technologies that have been applied to improve learning since they seek for the personalization and automation of this process. The literature reports important works, most of them focused on the instructional process or addressed to the virtual educational resource delivery. However, as a reflection of the traditional education situation, the assessment component and more specifically the adaptive assessment have not been given the same relevance within the learning process. Based on the premise that learning assessment is a fundamental component, which seeks to provide feedback from learning outcomes, it is proposed an assessment system that adapts to the specific conditions of each student. To do so, the system is able to recognize the academic differences and the learning style of the apprentices by modeling through a Multi-Agent System. In addition, this approach takes benefit from the great distribution advantages granted by this artificial intelligence technique. The preliminary validation of the system shows the possibilities that are open by using this kind of approaches.

Keywords: Adaptive assessment · Multi-Agent Systems · Virtual education

1 Introduction

Information and Communication Technologies (ICT) not only have allowed the development of all modern activities but also have generated a relevant impact on teaching and learning processes by raising new paradigms. From an educational point of view, curricular planning includes following issues: (1) the definition of educational goals or competencies to achieve, (2) the determination of the activities that allow these goals to be reached, (3) educational resources of different types that support learning activities and (4) the learning assessment. The latter one is understood as a systematic and permanent process of validating the whole teaching-learning process. From a methodological point of view, the assessment process can be seen as the result of the comparison between observed and expected learning situations [1].

R. Gennari et al. (Eds.): MIS4TEL 2019, AISC 1007, pp. 164–172, 2020.
https://doi.org/10.1007/978-3-030-23990-9_20

Technological advances, especially Artificial Intelligence (AI) techniques, have incorporated characteristics of flexibility and adaptability in teaching-learning environments. Although, there are several components involved in an educational virtual system, as shown in Fig. 1, not all the efforts and benefits have been addressed to the instructional process. In addition, these systems do not select appropriate digital content based on permanent and non-permanent characteristics of the students that could be used in the assessment process where general tests are often used for all students. The importance of the evaluation is that it plays a relevant role in the learning process, because the knowledge acquired by learners are validated and the shortcomings and/or strengths from students are detected [2].

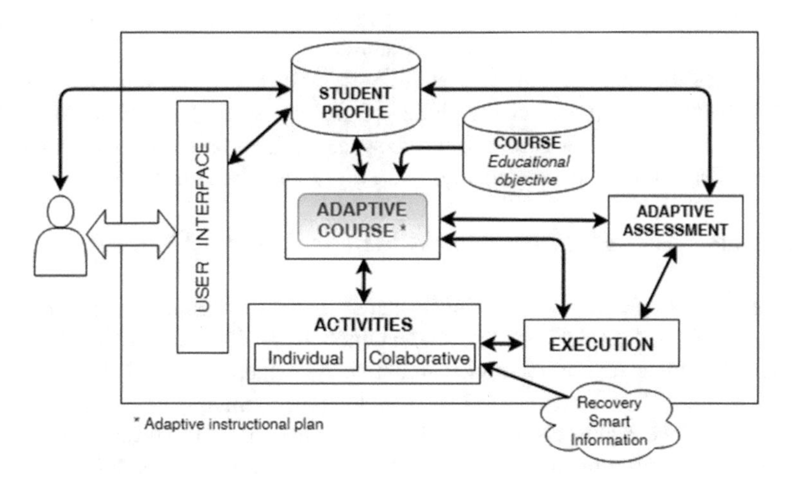

Fig. 1. Blocks in an environment of adaptive courses [3].

It is important to highlight that the development of virtual courses up to now has had a very marked boom towards the development of digital contents. However, the assessment process has been handled as a transposition of the face-to-face assessment with pencil and paper, without considering the new possibilities offered by artificial intelligence and adaptive systems. The advances in Artificial Intelligence techniques and its application to the adaptive systems open new possibilities that the researchers have not already completely exploited and now become an area of special interest for its development and implementation in an automated way. Relevant papers and reviews concerning this topic can be found in [4–9]. The adaptive assessment is a challenge that implying to create a personalized test, that includes the most suitable activities of evaluation for a specific student, taking into consideration the particular learning characteristics, needs and ability [10], or the new demands to knowledge, skills and attitudes [11].

The design and construction of personalized assessments taking advantage of adaptive techniques involve the phases of selection of achievements to be assessed, definition and recovery of the student profile, determination of the learning items, course grading and feedback to the students about their results. To achieve a real adaptability, the tests must be fixed according to an adaptation strategy, even allowing for the change of the goal to be assessed, looking for concrete evidences about the shortcomings or even modifying the difficulty degree to determine the level obtained by the student.

Recent research works such as the one presented in [12] define some metrics related to different aspects of the learning assessment process in a virtual environment and supported by intelligent agents.

The aim of this paper is to propose and present the most suitable characteristics of an assessment process that adapts to the academic conditions and psycho pedagogical preferences of the students. The process of analysis, design, and construction of the personalization module of the assessment is presented. This module is part of an experimental platform of adaptive courses and has been modeled taking advantage of the possibilities of the intelligent agent systems.

The rest of the paper is organized as follows: while Sect. 2 presents the conceptual proposal, Sect. 3 quickly shows the phases of analysis and design of the Multi-Agent System. Next, in Sect. 4, it is exhibited the system validation carried out in a blended learning undergraduate course. Finally, Sect. 5 presents the conclusions.

2 Proposal for Adaptive Assessment in Virtual Courses

Based on the ITS (Intelligent Tutorial Systems) model this proposal concepts such as domain model, pedagogical model, student model, the interface model [13] and is complemented by an additional model: Adaptive assessment. Figure 2 shows the whole structure of the system. Notice as in Fig. 2, all the other components of the system must be involved, which highlights the relevance of this activity in the process. In this way, it is proposed an assessment process that adapts to the academic conditions and to the psycho pedagogical characteristics and preferences of the students.

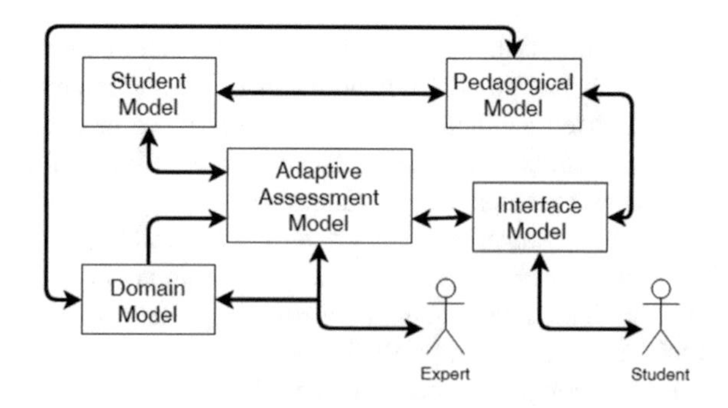

Fig. 2. Proposed system architecture components

According to the approach proposed in [14], the learning assessment must be guided by the Educational Objectives (EO) proposed to the virtual course, which the student is expected to achieve. In fact, the EO are the guide of the whole learning process and the assessment stage is a privileged moment for the negotiation of the pedagogical contract between the teacher and the students in the class [15].

As mentioned before the assessment stage starts from the definition of EO and therefore must reflect the proposed structure for them. According to the vision presented in [16], the structure of the course is hierarchical and the EO are defined to a fine level of detail. The process must identify where the student has shortcomings, for which it begins with the assessment of the objectives of higher order. Then, the process descends by considering the subjective objectives to locate exactly where the students have difficulties. This would allow for re-planning the learning tasks in this concrete case. At this point, it should also be considered the issue about the difficulty level of questions. The Questions Bank must respect the rules for this type of repositories and metadata is included for each question that associates it with the EOs that they assess.

In addition, the adaptation of the assessment is guided by the EO hierarchical structure, by the level of difficulty associated with them, and by the type of activities with which it is assessed. The activities are selected according to the student's learning style, which is an innovative approach to the proposal and takes advantage of previous works aimed at the process of delivering educational materials. For the selection of the type of activity, the proposal presented in [17] is applied.

3 MAS Proposal for Adaptive Assessment

The learning environments have received the influence of artificial intelligence and mark an important trend at present. These environments seek to improve the ability of continuous adaptation of the different phases of the educational process by using the characteristics and/or preferences of the students.

A Multi-Agent Systems (MAS) is an organized society composed of semi-autonomous agents that interact with each other, either to collaborate in the solution of a set of problems or in the achievement of a series of individual or collective goals. The MAS principles have shown adequate potential in developing learning systems since the nature of teaching-learning problems is more easily faced through a cooperative approach. However, much remains to be done in this field, as mentioned in [18]: "While our understanding of learning agents and multi-agent systems has advanced significantly, most applications are still simple on scaled-down domains, and, in fact, most methods do not scale up to the real world".

The development methodology adopted was MASCommonKADS, which is an extension of CommonKADS, with aspects that are relevant to MAS and integrating techniques of object-oriented methodologies in order to facilitate their use. The methodology states seven models for the definition of development, some of which are described below. Figure 3 summarizes the determined agents and the relationship established between them. Next, the student and adaptation agents are detailed which represent the basis of the personalization system.

Student Agent

The student agent is composed of the class called *Aestudiante* that has a cyclical behavior and a method that separates chains. It waits for the arrival of a message, when it receives the message it is divided into three parts and begins to ask who sent the message and what is the name of the petition or the information that is being sent. Then, based on this information performs the following operations:

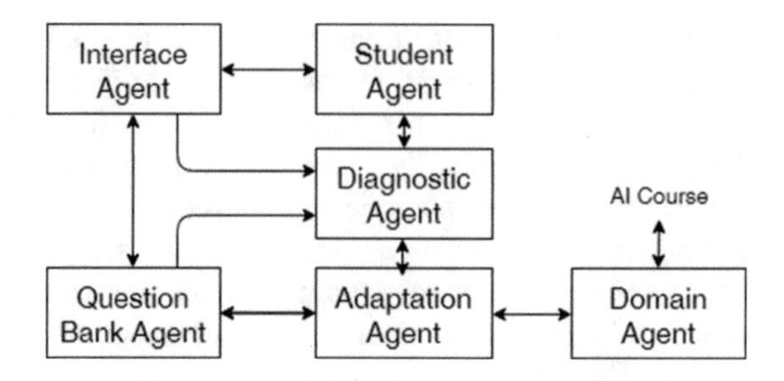

Fig. 3. Proposed system components

- Delivery of the student profile to the adaptation agent.
- If the student does not have a profile, it makes the profile request to the diagnostic agent. Message: ID | LS | PT, ConversationId ("perfil");
- It stores the assessment result of an educational objective (EO). In addition, if the EO is accomplished or not, with which question, and what is receiving from the diagnostic agent. Message: ID | OE | FULFILLED | IDP, ConversationId ("objeducativo");
- Responds to the adaptation agent if a certain EO was fulfilled or not.

Adaptation Agent

The *adaptation agent* is composed of the class called *Aadaptacion* that has a cyclic behavior, and the following methods: Method that separates chains, method that asks for children of an EO, method that receives the children of an EO, method that asks if an EO has been achieved, method that asks questions (learning objects) for an EO and a student profile. The adaptation agent is waiting for the arrival of a message, which can be an ACL text message or an object. If the message is ACL text it is divided into three parts and begins to ask who sent the message and what is the name of the request or information that is being sent. Then, based on this information the adaptation agent performs the following operations:

- Receive assessment request from an external agent.
  Message: ID | EO, ConversationId ("EO");
- Request the profile of a student.
- Ask questions for a specific profile and EO to the base agent of questions
  Message: EO | LS | PT, ConversationId ("preguntas");
- Receive questions. Message: vector with the questions, ConversationId ("preguntas");
- Receives response from an EO by the diagnostic agent.
  Message: ID | EO | CUMPLIO, ConversationId ("respuesta");
- Requests children of an EO to the domain agent. Message: EO, ConversationId ("hijosdeEO");
- Receive children. Message: chain with the children, ConversationId ("hijosdeEO");
- Ask to the diagnostic agent whether or not an EO was accomplished.

Figure 4 shows the steps needed to follow in order to deliver the personalized assessment to the student. The process starts with the request of an external agent so that the selection process of an assessment of a given student (adaptation agent) begins. After collecting the data of the student profile (student agent), the system sends a question request to the question bank agent. This request is selected in the question bank database according to the following criteria: EO to be assessed, student's learning style, and technological profile. After the question has been shown to the student and he/she gives a response (interface agent), the result is verified whether the EO is fulfilled. Otherwise, a re-adaptation process is initiated to detect where there are flaws in the process, which have not allowed the expected output.

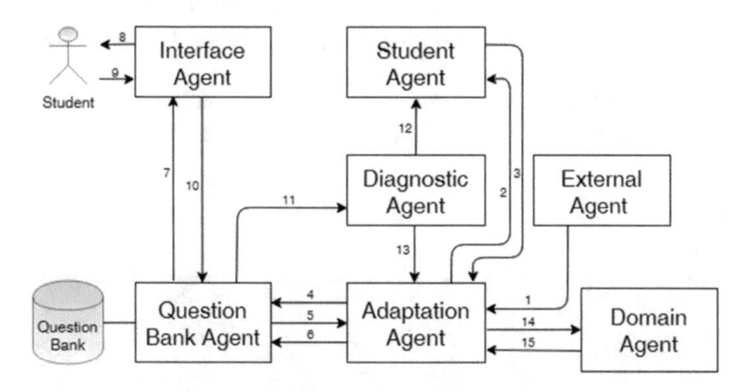

Fig. 4. Sequence of activities in the MAS.

JADE (Java Agent Development Framework) that is a FIPA-compliant framework for public use was employed to develop the assessment agents system. The MySQL database system stores the knowledge base and the structure of the virtual course. Figure 5 exhibits the sequence diagram of the system. It can be noticed the different relationships between the agents and the actions needed to ensure that the assessment is adapted to each student according to their characteristics.

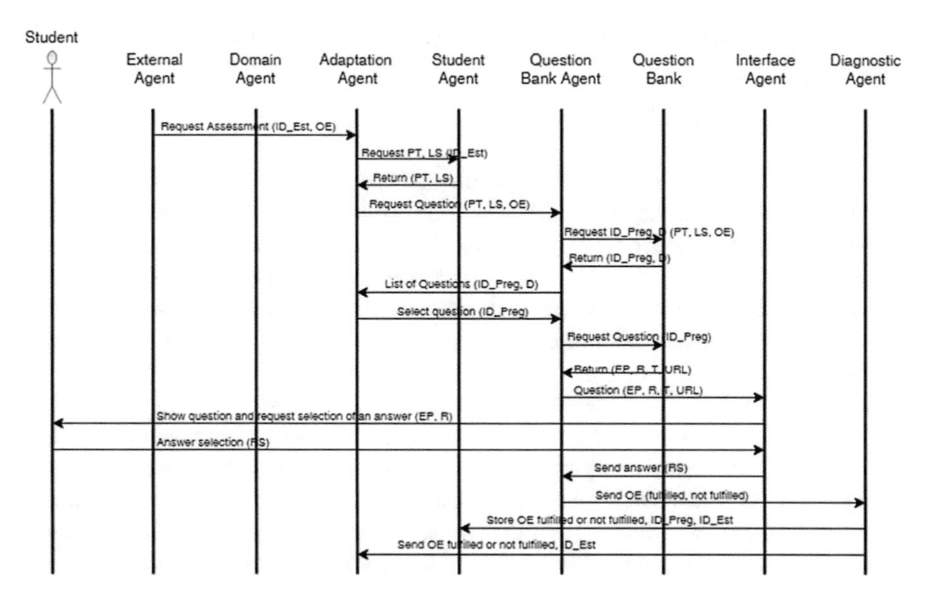

Fig. 5. Sequence diagram of the proposed system

4 Proposal Validation

The validation of the proposal was carried out using a blended undergraduate virtual course of audit in the chapter of computer-assisted auditing techniques. As a first step, the Fleming test was performed to determine the learning style of the students. A pair of EO concerning the topic to assess were defined which compose a hierarchical structure with higher levels of achievement. This structure that is decomposed into several lower levels as the designer would consider.

For each EO at the different levels, learning activities and objects were constructed based on different learning styles by following the patterns taken from [17], as shown in Fig. 6.

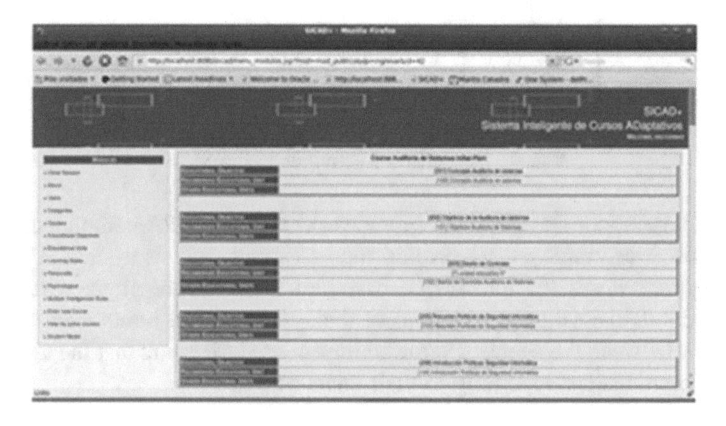

Fig. 6. Customized assessment-learning resources to deliver.

5 Conclusions and Future Work

The difficulty in modeling real-world problems in knowledge-based systems is manifest and only the skill to represent the system can ensure that the power of technologies becomes a real tool for everyday applications. Smart agents can act on behalf of users or entities involved in a system but in order to obtain satisfactory results depend on a rigorous application of the adopted methodology. The distribution of intelligence and its capacity for modularization are desirable characteristics of growing systems with tuning needs.

The way in which the model is presented allows for choosing different alternatives to carry out the assessment process adaptation, which is presented as a neutral tool with respect to the pedagogical thoughts and learning styles of those who wish to implement it. At present, the interactions have been tested and the results evaluated, with a view to the integration process within the virtual courses experimental platform. In the tests of question selection, the proposal has been validated according to the expected results.

Acknowledgments. The research was partially funded by the project "Fortalecimiento docente desde la alfabetización mediática Informacional y la CTel, como estrategia didáctico-pedagógica y soporte para la recuperación de la confianza del tejido social afectado por el conflicto" code SIGP 58950 of the program "Reconstrucción del tejido social en zonas de pos-conflicto en Colombia" whit code SIGP 57579 supported by Fondo Nacional de Financiamiento para la Ciencia, la Tecnología y la Innovación, Fondo Francisco José de Caldas whith contract 213-2018.

References

1. Gagne, R.M., Briggs, L.J., Wage, W.: Instructional Design. Rinehart and Winston Inc, New York (1988)
2. Salazar, O., Ovalle, D.A., de la Prieta, F.: Towards an adaptive and personalized assessment model based on ontologies, context and collaborative filtering. In: 15th International Conference on Distributed Computing and Artificial Intelligence, DCAI (2019)
3. Duque, N.: Modelo Adaptativo Multi-Agente para la Planificación y Ejecución de Cursos Virtuales Personalizados, Universidad Nacional de Colombia - Sede Medellín (2009)
4. Wang, T.H.: Developing web-based assessment strategies for facilitating junior high school students to perform self-regulated learning in an e-Learning environment. Comput. Educ. **57**(2), 1801–1812 (2011)
5. Neill, J.: Performance Assessment in Online Learning, pp. 1–5 (2005)
6. Nacheva-Skopalik, L., Green, S.: Adaptable Personal E-Assessment. Int. J. Web-Based Learn. Teach. Technol. **7**(4), 29–39 (2012)
7. Conrad, S., Clarke-Midura, J., Klopfer, E.: A framework for structuring learning assessment in a massively multiplayer online educational game: experiment centered design. Int. J. Game-Based Learn. **4**(1), 37–59 (2014)
8. Nakatsuji, M., Fujiwara, Y.: Linked taxonomies to capture users' subjective assessments of items to facilitate accurate collaborative filtering. Artif. Intell. **207**, 52–68 (2014)
9. El Alami, M., Maroc, T., De Arriaga, F.: Fuzzy assessment for affective and cognitive computing in intelligent e-Learning systems. Int. J. Comput. Appl. **100**(10), 40–46 (2014)
10. Chrysafiadi, K., Troussas, C., Virvou, M.: A framework for creating automated online adaptive tests using multiple-criteria decision analysis. In: Proceedings - 2018 IEEE International Conference on Systems, Man, and Cybernetics, SMC 2018 (2018)
11. Krouska, A., Troussas, C., Virvou, M.: Computerized adaptive assessment using accumulative learning activities based on revised bloom's taxonomy. In: 12th Joint Conference on Knowledge-Based Software Engineering, JCKBSE 2018 (2019)
12. Salazar, O.M., Ovalle, D.A., Duque, N.D.: Evaluation of metrics-based performance of a ubiquitous e-learning multi-agent context-aware system, using ontologies|Evaluación del desempeño basado en métricas de un sistema pedagógico multi-agente, ubicuo sensible al contexto y apoyado en ontologías, Form. Univ., vol. 9, no. 3 (2016)
13. Latham, A., Crockett, K., McLean, D., Edmonds, B.: A conversational intelligent tutoring system to automatically predict learning styles. Comput. Educ. **59**, 95–109 (2012)
14. Díaz, L.: Los Objetivos Educacionales: Criterios Claves para la Evaluación del Aprendizaje. Universidad de Puerto Rico, Río Piedras (1994)
15. Morrissette, J.: Modos de interacción como fundamento en la realización de una evaluación formativa no instrumentada. Estud. pedagógicos **41**(2), 373–388 (2015)

16. Duque, N., Ovalle, D.: Artificial intelligence planning techniques for adaptive virtual course construction. Rev. DYNA **78**, 70–78 (2011)
17. Duque-Mendéz, N., Tabares Morales, V., Vicari, R.: Learning Object Metadata Mapping with Learning Styles as a Strategy for Improving Usability of Educational Resource Repositories, Rev. Iberoam. Tecnol. del Aprendiz., **11**(2), pp. 101–106 (2016)
18. Kudenko, D., Kazakov, D., Alonso, E. (ed.) Adaptive Agents and Multi-Agent Systems III. Adaptation and Multi-Agent Learning, vol. 4865 (2008)

Author Index

R. Gennari et al. (Eds.): MIS4TEL 2019, AISC 1007, pp. 173–174, 2020.
https://doi.org/10.1007/978-3-030-23990-9

Printed in the United States
By Bookmasters